THE AMERICAN CIVIL WAR: THEN AND NOW

THE AMERICAN CIVIL WAR: THEN AND NOW

G.R WOODWARD

New York Madrid London

ISBN:978-1-956768-33-6

FIERO PUBLISHING COMPANY

DEDICATION

To my dog Skippy, who has made my life more enjoyable every moment and who has spent adequate time interrupting my efforts in writing this piece on the American Civil War.

CONTENTS

1	Origins	11
2	Public Opinion	22
3	Life in America	30
4	President Lincoln's Goals	38
5	The Opposition	45
6	Battles Fought	58
7	Military Strategy	70
8	Arsenal of Weapons	82
9	Mass Media	86
10	War Effort	95
11	Slavery	103
12	Social Life	121
13	Economics and Finance	140
14	Freedom and Reconstruction	159
15	Segregation	174
16	The Southern Promise	185
17	Hate Groups	205
18	A Desire For Strong Leadership	216

1
ORIGINS

The American Civil War, which took place from 1861 to 1865, was primarily fought over the issue of slavery. The Southern states wanted to preserve the institution of slavery and maintain their agrarian economy and way of life, while the Northern states were increasingly opposed to slavery and sought to abolish it. Tensions continued to escalate, leading to the secession of eleven Southern states and the formation of the Confederate States of America. The war ultimately erupted when Confederate forces attacked Union-held Fort Sumter. While slavery was the main catalyst, other factors, such as states' rights, economic differences, and political disagreements, also contributed to the outbreak of the conflict.

During the Battle of Fort Sumter, which took place from April 12-14, 1861, there were no recorded deaths among the Union defenders of the fort. However, there were casualties among the soldiers involved in the fighting.

The Confederate forces, commanded by General P.G.T. Beauregard, bombarded Fort Sumter, located in Charleston Harbor, South Carolina. The bombardment caused destruction and injuries among the Union soldiers present. According to historical records, there were five Union soldiers wounded during the battle.

It is important to note that the Battle of Fort Sumter was the starting point of the American Civil War. While there were no fatalities in this particular battle, the conflict as a whole would go on to result in the deaths of hundreds of thousands of soldiers on both sides over the course of four years.

The Battle of Fort Sumter resulted in a Confederate victory. After a two-day bombardment by Confederate forces, the Union garrison at Fort

Sumter, commanded by Major Robert Anderson, surrendered on April 13, 1861. The Confederate forces, under the command of General P.G.T. Beauregard, took control of the fort.

While the Union defenders of Fort Sumter did not have any deaths during the battle, they were compelled to evacuate and surrender due to the damage inflicted upon the fort and the threat of further bombardment. The Confederate victory at Fort Sumter marked the beginning of the American Civil War.

During the American Civil War, the Confederacy sought foreign recognition and support to bolster its cause against the Union. While the Confederacy hoped to secure England's support, they were ultimately not recognized as an official ally by the British government. However, there were several factors that led some in England to sympathize with the Confederacy:

England was a major manufacturing center during this time and depended heavily on Southern cotton for its textile industry. The

Confederacy believed that by stopping the supply of cotton to England, it could pressure the British government into recognizing its independence. This economic factor made some British businessmen and politicians sympathetic to the Confederacy.

The Union implemented a naval blockade of Southern ports in an attempt to cut off Confederate trade. This blockade affected British merchants who relied on Southern cotton imports. Some in England saw this blockade as an infringement on their trade and opposed the Union's actions, leading to sympathy for the Confederacy.

There were longstanding ties between the British aristocracy and Southern plantation owners. Many wealthy plantation owners in the South had British connections and family ties, resulting in empathy and support for the Confederate cause among the British elite.

Some British politicians saw potential advantages in supporting the Confederacy. They thought that recognizing the Confederacy could weaken the United States, which at the time was an emerging global power and potential rival to British interests.

Southern culture played a significant role in shaping the identity of the Confederate states during the American Civil War. The Confederate states, which were primarily located in the Southern region of the United States, had distinct cultural attributes that helped define and reinforce their identity as a separate nation.

The Southern economy was primarily based on agriculture, with cotton being the dominant cash crop. The plantation system, which relied heavily on enslaved labor, was central to Southern society. This agrarian lifestyle and reliance on agriculture shaped the economic and social structure of the Southern states and influenced their collective identity.

The concept of honor was deeply ingrained in Southern culture. Southerners valued notions of personal honor, family integrity, and a code of behavior rooted in chivalry. These ideals influenced the conduct of Southern soldiers in battle and the society at large. The defense of personal and familial honor was seen as an integral part of Southern identity.

Slavery was a fundamental component of the Southern economy and society. Southern culture developed around the institution of slavery, with slaveholding seen as a marker of status and power. The Confederacy's identity was largely shaped by the defense of slavery and the belief in white supremacy as a core value.

Southerners, even before the Civil War, had a strong sense of regional identity and pride. They viewed themselves as distinct from the Northern states in terms of history, values, and way of life. This regional pride was reinforced by literature, music, art, and traditions celebrating the unique qualities of the Southern states.

The Southern states had a strong military tradition, with many Southerners serving in the armed forces. This tradition of military service, coupled with the notion of honor and bravery, became intertwined with Southern culture and played a significant role in the Confederacy's identity during the war.

The aforementioned aspects of Southern culture, among others, contributed to the development of a distinctive Southern identity that helped shape the Confederacy. They provided a framework for the Confederate states to unite around common values, beliefs, and goals during the Civil War.

Religion played a significant role during the American Civil War. Both the Union and Confederate sides drew on religious beliefs and ideologies to justify their respective causes and as a source of inspiration for their soldiers.

Religious rhetoric and interpretations were used to justify the causes of both the Union and Confederate sides. Many Northern religious leaders saw the war as a divine punishment for the sin of slavery, emphasizing the abolitionist cause. In contrast, Southern religious leaders often defended slavery as biblically sanctioned, viewing the war as a defense of their way of life and their ability to practice their religion.

Soldiers and civilians on both sides relied heavily on prayer and religious faith for strength and comfort during the war. Chaplains were an integral part of the military, offering religious services, pastoral care, and moral support to soldiers.

Religious teachings and principles were invoked to provide ethical guidance for soldiers and officers. Religious leaders on both sides encouraged soldiers to fight with honor and courage, offering moral justifications for their actions on the battlefield.

Many soldiers and civilians believed in divine intervention in the outcome of the war. Prayers, sermons, and religious rituals often included pleas for God's favor and assistance.

Religion provided solace and hope for individuals and communities affected by the war. The belief in an afterlife and the promise of divine justice offered comfort in the face of loss and suffering.

Religious organizations and individuals played a vital role in providing support and relief to soldiers and civilians during the war. They established hospitals, offered medical assistance, and provided care for the wounded and displaced.

It is important to note that religion during the Civil War was complex and not monolithic. Different religious denominations and individuals held various perspectives on the war and its causes. Religion both unified and divided communities, and interpretations of religious texts and beliefs varied widely across the nation. The role of religion during the Civil War is a topic of ongoing scholarship and analysis.

Public opinion during the Civil War was divided and complex, reflecting the deep divisions within the country over the issues of slavery, states' rights, and secession.

In the northern states, there was a general belief in the preservation of the Union and opposition to secession. However, this did not necessarily mean there was unified support for the abolition of slavery. Many northern states had strong economic ties to the institution of slavery, and some were hesitant to take a hard stance against it.

Over time, as the war progressed and the Emancipation Proclamation was issued in 1862, public opinion in the North shifted more towards anti-slavery sentiment. Popular support for the Union cause grew, though it was not without dissent and opposition, particularly from Copperheads, who were Democrats advocating for a negotiated peace with the Confederacy.

In the southern states that seceded from the Union, public opinion was overwhelmingly supportive of the Confederate cause. Many southerners saw the defense of their states' rights and the institution of slavery as justifications for secession. The Confederate government and its leaders rallied public support and emphasized the importance of preserving their way of life and their right to self-governance.

Slaveholders, in particular, held a vested interest in preserving slavery, and public opinion among white southerners generally favored its continuation. However, it's worth noting that not all white southerners were slaveholders or necessarily supported the Confederacy. Some were Unionists who opposed secession, and others were non-slaveholding southerners who were more economically aligned with the North.

2
PUBLIC OPINION

Public opinion among African Americans during the Civil War was also diverse. Enslaved individuals desired their freedom and actively sought ways to escape and join the Union Army or find refuge in Northern states. Free African Americans and abolitionists in the North were vocal in their support for the Union cause and the abolition of slavery.

One of the most famous books written during the Civil War is "Uncle Tom's Cabin" by Harriet Beecher Stowe. The novel, published in 1852, depicted the harsh realities of slavery and its impact on enslaved individuals. It played a significant role in shaping public opinion on slavery and fueling the abolitionist movement. "Uncle Tom's Cabin" is considered a powerful and influential piece of literature in American history.

Public sentiment during the Civil War was shaped by a range of factors, including regional and economic interests, political affiliations, personal experiences, and moral convictions. It is important to recognize that public opinion was not monolithic, and there was a spectrum of viewpoints within each region and group.

During the American Civil War, Confederate agents made several foreign trips in an attempt to garner support for the Confederate cause and secure assistance from foreign powers. These agents were often dispatched to Europe with various objectives, including lobbying for diplomatic recognition, seeking military aid, and promoting Confederate cotton exports. Some notable examples include:

1. James M. Mason and John Slidell: In 1861, James M. Mason and John Slidell were Confederate diplomats dispatched to Europe. They were captured by the Union Navy while on a British ship, the RMS Trent, which led to a diplomatic incident known as the Trent Affair. The incident strained

relations between the United States and the United Kingdom.

2. James D. Bulloch: Bulloch was a Confederate naval officer and agent who played a significant role in acquiring and converting ships for the Confederate Navy. He traveled to Europe, mainly in Britain, to negotiate contracts and procure warships, including the famous CSS Alabama.

3. Thomas Nelson Conrad: Conrad was another Confederate agent who traveled to Europe, mainly in France, seeking military and diplomatic support for the Confederacy. He worked to secure French recognition of the Confederacy and also explored the possibility of French intervention.

These trips by Confederate agents were part of an overall effort to gain international recognition, secure military aid, and undermine the Union's position diplomatically. While these missions had some limited successes, ultimately, foreign powers did not provide the level of support that the Confederacy needed to turn the tide of the war in their favor.

Despite these sympathies and considerations, the British government ultimately decided not to officially recognize the Confederacy. This was largely due to other factors, such as British public opinion and concerns about potential backlash from the Union. Furthermore, the Emancipation Proclamation issued by President Abraham Lincoln, which declared enslaved people free in Confederate-held territory, made it politically difficult for Britain to support a pro-slavery cause.

In the end, while England was not an official ally of the Confederacy, there were significant elements within England that sympathized with the Confederacy and were affected by the Civil War's impact on the cotton trade.

Union agents also made foreign trips during the Civil War to pursue various objectives related to diplomacy, intelligence, and economic matters. Some notable examples include:

1. Charles Francis Adams: Adams served as the United States Ambassador to the United Kingdom

during the Civil War. He played a crucial role in preventing British recognition of the Confederacy and securing British neutrality. Adams effectively presented the Union's case, highlighting the legality of the blockade and the moral and economic issues related to supporting the Confederacy.

2. William H. Seward: Seward, who served as Secretary of State under President Abraham Lincoln, sent diplomatic agents to Europe during the war. They worked to counter Confederate efforts at gaining recognition and sought to sway public opinion in favor of the Union cause.

3. John Bigelow: Bigelow was another Union diplomat who worked in France during the Civil War. He navigated delicate relationships with the French government, seeking to prevent French intervention on behalf of the Confederacy.

4. Union Naval Intelligence Agents: The Union Navy dispatched intelligence agents to Europe to gather information about Confederate activities, monitor European shipyards for warship

construction, and identify Confederate naval operations. These agents worked covertly to collect vital intelligence for the Union.

Overall, Union agents' foreign trips were aimed at diplomatic efforts, intelligence gathering, and economic negotiations to bolster the Union's position and counter Confederate efforts to gain international recognition and support.

Both Britain and France ultimately decided not to support the Confederacy during the American Civil War for a variety of reasons. Britain and France relied heavily on imports of raw materials, particularly cotton, from the Southern states. Initially, there were concerns that the Union blockade of Confederate ports would disrupt the supply of cotton, causing economic hardship in Europe. However, Britain and France were able to find alternative sources of cotton, mitigating these concerns and reducing their interest in supporting the Confederacy.

Britain and France were initially hesitant to intervene in the American Civil War due to their policy of neutrality. They viewed the conflict as an internal matter and were reluctant to take sides. Both countries recognized the Union blockade as a lawful act and sought to maintain diplomatic relations with the United States. The abolitionist movement had a significant impact on public opinion in Britain and France. The moral imperative to end slavery and the networks of anti-slavery activists influenced public sentiment. The Emancipation Proclamation further solidified the Union's position as an anti-slavery force, reinforcing the reluctance of European powers to support the Confederacy.

Supporting the Confederacy carried significant risks and potential costs for Britain and France. There were concerns about the potential consequences of openly supporting a secessionist movement against a unified United States. Additionally, both countries were recovering from previous conflicts, such as the Crimean War, and were not prepared to engage in another wide-scale military intervention. The Union made diplomatic efforts to dissuade British and French intervention. The threat of war with the United States, as well as strategic and economic incentives, led both countries to ultimately refrain from openly supporting the Confederacy.

While Britain and France did not formally support the Confederacy, there were instances of unofficial aid, such as the construction of Confederate warships in British shipyards. These actions strained relations between the United States and these European powers. Overall, economic considerations, political neutrality, public opinion, risks and costs, and Union diplomatic efforts all played a role in influencing Britain and France to abstain from supporting the Confederacy during the American Civil War.

3
LIFE IN AMERICA

Life in America during the 1860s was deeply affected by the American Civil War, which began in 1861 and ended in 1865. The war had a significant impact on various aspects of daily life, economy, politics, and society.

The Civil War between the Union (Northern states) and the Confederacy (Southern states) dominated the decade. Families and communities were divided, with individuals choosing sides and often fighting against friends and relatives. The war brought immense destruction, loss of life, and economic disruption to the nation.

The 1860s saw intense debates and conflicts over slavery, states' rights, and the future direction of the nation. These issues led to the secession of several Southern states and the formation of the

Confederacy. Tensions between the North and the South were high, and the war brought about profound social and political changes.

The Civil War resulted in a significant number of men serving in the military. Families and communities experienced the anguish of seeing loved ones enlist, go off to war, and face the hardships and dangers of combat. The war also caused immense casualties, with hundreds of thousands of soldiers losing their lives or suffering injuries.

The war had major economic consequences. Agricultural states in the South faced severe disruption as enslaved individuals left plantations, and crops were abandoned or destroyed. The North, on the other hand, experienced rapid industrialization and economic growth due to increased production to support the war.

The Civil War played a pivotal role in the abolition of slavery. President Abraham Lincoln's Emancipation Proclamation, issued in 1863, declared enslaved individuals in Confederate states to be free. The war ultimately led to the passage of the 13th Amendment in 1865, officially abolishing slavery throughout the United States.

The war affected women's lives significantly, as they took on new responsibilities and played crucial roles on both the home front and the battlefront. Women in the North and the South worked as nurses, teachers, and even spies. The war accelerated the women's suffrage movement, as women started asserting their rights and demanding greater participation in public life.

Women played vital roles during the Civil War, both on the home front and in various capacities directly related to the conflict. Here are some of the activities and roles women undertook during the war:

Many women served as nurses, both in military hospitals and on the battlefield, tending to wounded soldiers from both the Union and Confederate sides. Notable figures like Clara Barton and Dorothea Dix made significant contributions in organizing medical care and improving conditions for the wounded.

Women served as spies and intelligence agents for both the Union and Confederate forces. They often used their social connections and ability to blend in to gather and relay valuable information about troop movements, supply lines, and enemy plans.

While it was uncommon, a small number of women disguised themselves as men and fought as soldiers in the Civil War. Examples include Sarah Emma Edmonds, who served as a Union soldier, and Loreta Janeta Velázquez, who fought for the Confederacy.

Women participated in various relief efforts, both organized and informal, to provide aid and support to soldiers and their families. They sewed uniforms and blankets, collected supplies, raised funds, and operated aid societies to assist soldiers and their communities affected by the war.

With many men serving in the military, women took on more responsibility for managing farms and plantations, as well as working in factories and other labor-intensive industries to support the war effort. Unfortunately, prostitution also became more prevalent during the war, with some women turning to this profession to survive or to follow the armies. Brothels often sprang up around military camps and served soldiers from both sides.

Women were instrumental in forming and supporting organizations like the United States Sanitary Commission, which aimed to improve medical care for soldiers and veterans. They raised funds, collected supplies, and worked to ensure the health and well-being of soldiers.

It's important to note that the experiences and contributions of women during the Civil War varied greatly depending on their geographical location, social class, and race. African American women, for example, faced additional challenges and often supported the Union cause by serving as nurses, laundresses, or workers in contraband camps.

Women played diverse and essential roles during the Civil War, from combat support to homefront activities, contributing significantly to the war effort and the social changes that followed. Towards the end of the 1860s, the nation entered a challenging phase known as Reconstruction. This period focused on rebuilding the South, reintegrating the former Confederate states into the Union, and addressing issues of civil rights and equality for newly freed African Americans.

Generally speaking, life during the 1860s in America was marked by the upheaval and devastation of the Civil War. It intensified existing social and political divisions while paving the way for significant changes in American society.

The specific design of the Confederate flag, commonly known as the "Southern Cross" or Confederate flag, went through several iterations during the American Civil War. The Confederate government adopted its first official flag, known as the "Stars and Bars," on March 4, 1861. This flag featured three horizontal stripes: two red stripes at the top and bottom, with a thick white stripe in the middle. In the upper left corner, there were seven white stars on a blue background, representing the seven seceded Southern states.

However, the "Stars and Bars" design proved problematic due to its similarity to the Union flag, causing confusion on the battlefield. Consequently, there was a need for a new design that would be more recognizable. A battle flag, featuring a blue saltire outlined in white with 13 white stars, was introduced as a distinct Confederate symbol.

The battle flag, often referred to as the "Confederate flag," has become one of the most recognizable symbols associated with the Confederacy. It was never officially adopted as the national flag of the Confederacy, but it gained prominence as a military flag and was widely used as a symbol by Confederate soldiers and units.

It is important to note that the Confederate flag, in its various designs, has undergone different interpretations and regained popularity during various periods in American history, including during the Civil Rights Movement of the 1960s. Its usage continues to be a subject of debate and controversy in the present day.

4 PRESIDENT LINCOLN'S GOALS

Lincoln's primary goal was to preserve the United States as a unified nation. He believed that the secession of Southern states was unconstitutional and violated the principles of the American democracy. Lincoln was determined to prevent the Confederacy from becoming an independent nation and maintain the integrity of the Union.

While initially Lincoln's main objective was to preserve the Union rather than abolish slavery, as the war progressed, his views evolved. He recognized that slavery was a central cause of the conflict and believed that ending slavery was necessary for a more just and equitable society. In 1863, he issued the Emancipation Proclamation, declaring enslaved people in Confederate-held territories free, and ultimately supported the passage of the Thirteenth Amendment that abolished slavery.

The Civil War led to widespread violence and disruption across the nation. Lincoln was committed to restoring order and stability within the Union. He sought to bring an end to hostilities, mend the divisions between Northern and Southern states, and establish a lasting peace through reconciliation and reconstruction efforts. Lincoln firmly believed in the principles of democracy and the rule of law. He aimed to defend and uphold the democratic ideals upon which the country was founded. He emphasized the importance of unity, the peaceful transfer of power, and the rights and liberties of citizens.

Overall, Lincoln's primary goals were the preservation of the Union, the abolition of slavery, restoration of order, and the defense of democratic principles. His leadership and determination during the war have solidified his place as one of the most influential presidents in American history.

Lincoln's presidency faced numerous challenges during the American Civil War. Here are some key challenges: The Southern states' secession and formation of the Confederate States of America posed a significant challenge to Lincoln's presidency. The Confederacy's attempt to establish itself as an independent nation threatened the unity of the United States and challenged Lincoln's authority.

Lincoln had to navigate the complexities of leading the Union army and formulating effective military strategies. He faced the challenge of finding capable military commanders who could effectively lead and coordinate Union forces. As the war progressed, Lincoln had to make difficult decisions regarding tactics, resource allocation, and personnel changes.

Lincoln faced criticism and differing opinions from various factions within the Union itself. Some sections of the population opposed the war effort and challenged his policies, particularly on issues such as the suspension of habeas corpus and the draft. Building and maintaining public support for the war was a continuous challenge for Lincoln.

During the Civil War era in the United States, the country's demographics were characterized by a growing population and distinct regional differences. The overall population of the United States was increasing rapidly during the mid-19th century. In 1860, the population was approximately 31 million people, with significant growth compared to earlier decades.

The United States was divided into two main regions with contrasting demographics. The Northern states had a higher population density, primarily due to industrialization and immigration. The urban centers in the North, such as New York, Philadelphia, and Boston, were growing rapidly. The Southern states had a more rural and agrarian society. The slaveholding states of the South, known as the "Old South," were heavily reliant on an agrarian economy based on cotton and plantation agriculture.

Slavery was a significant demographic factor before and during the Civil War. In the Southern states, nearly four million enslaved individuals comprised a substantial portion of the population. Slavery was deeply intertwined with the region's social, political, and economic systems.

The majority of the population in the United States during the Civil War era consisted of people of European descent, primarily from British, German, and Irish backgrounds. African Americans constituted a significant minority, primarily concentrated in the Southern states.

The mid-19th century also witnessed significant waves of immigration, particularly to the Northern states. Immigrants from Ireland, Germany, and other European countries contributed to the nation's growing diversity, labor force, and urbanization. It's worth noting that these demographic factors played a role in shaping the political, social, and economic conflicts that culminated in the Civil War. The divide between the Northern and Southern regions, fueled by differences in demographics, economies, and attitudes toward slavery, became a central issue leading to the war.

The war exacerbated political divisions within the Union. Lincoln faced opposition from factions within his own Republican Party, such as the Radical Republicans, who pushed for more aggressive measures on issues such as emancipation and Reconstruction. Managing party divisions while striving for unity was a persistent challenge for Lincoln.

As Lincoln increasingly embraced the goal of ending slavery, he faced challenges related to race relations and the role of formerly enslaved people in society. Balancing the aspirations for equality and freedom with the complexities of Reconstruction and the integration of formerly Confederate states posed significant challenges.

Tragically, Lincoln was assassinated in April 1865, just a few days after the end of the Civil War. His untimely death meant that he did not witness the challenges of Reconstruction and ensuring a smooth transition for the nation after the war. His assassination also impacted how he would be remembered and the legacy he left behind.

Despite these challenges, Lincoln's leadership, steadfastness, and his ability to address these issues ultimately played a crucial role in the preservation of the Union and the abolition of slavery.

5
THE OPPOSITION

During the American Civil War, Jefferson Davis, as the president of the Confederate States of America, challenged Abraham Lincoln's presidency and the Union government in several ways. As the leader of the secessionist Confederate government, Davis challenged Lincoln's authority by actively opposing and seeking to establish a separate nation from the United States. The Confederate forces, under Davis's leadership, waged war against the Union.

Davis played a key role in formulating the military strategy for the Confederacy. He appointed several generals, such as Robert E. Lee, and provided overall guidance to the Confederate military. Through military action, Davis aimed to challenge and defeat the Union forces led by Lincoln.

Davis made diplomatic efforts to gain recognition and support for the Confederacy from European powers, such as Britain and France. By seeking diplomatic recognition, Davis aimed to undermine Lincoln's efforts to maintain the unity of the United States.

As the president of the Confederacy, Davis was seen as a political rival to Lincoln. He led the Confederate government, which sought to preserve slavery and states' rights, while Lincoln and the Union government viewed the preservation of the Union and the end of slavery as their primary goals. This ideological divide between Davis and Lincoln fueled their rivalry.

Ultimately, Jefferson Davis challenged Abraham Lincoln by leading the Confederate government and military to oppose and challenge the Union and its leadership during the American Civil War.

During the American Civil War, the number of soldiers who fought on the Confederate side and the Union side varied over the course of the conflict. Here are some estimates of the approximate number of soldiers involved:

Confederate Side:
- The Confederate army had an estimated total enlistment of around 800,000 to 1.5 million soldiers throughout the war. However, it is important to note that accurate figures are challenging to determine due to incomplete records and difficulties in tracking enlistments.

Union Side:
- The Union army saw enlistment numbers ranging from around 2.1 million to 2.8 million soldiers. As with the Confederate side, obtaining precise figures is challenging due to incomplete records and the presence of duplicate enlistments.

It is worth noting that these numbers represent total enlistments rather than the number of active soldiers at a given time. The total strength of the armies varied throughout the war due to factors such as casualties, desertions, and discharges. The Union had a substantial advantage in terms of manpower, which played a significant role in the outcome of the war. The larger population of the Northern states, coupled with higher rates of industrialization and immigration, enabled the Union to maintain a larger army and sustain a longer war effort.

It is also important to consider the diverse composition of both armies, with soldiers hailing from a range of backgrounds, including immigrants, free African Americans, and Native Americans, among others. The war had a profound impact on the lives of those who fought and the direction of the nation's history. During the American Civil War, both the Union and Confederate forces had African Americans serving in various capacities. However, there were significant differences in their roles and numbers.

Union Forces:

The Union Army actively recruited African American men to serve in the military. They formed segregated units known as the United States Colored Troops (USCT). By the end of the war, almost 180,000 African American soldiers and sailors had served in the Union forces. Initially, African American soldiers were paid less than their white counterparts, but this discrepancy was eventually rectified.

Confederate Forces:
Initially, the Confederate leadership was reluctant to allow African Americans to serve as soldiers due to a fear of slave rebellion and concerns about arming enslaved individuals. However, as the war progressed and the Confederacy faced manpower shortages, they began employing African Americans as impressed laborers, teamsters, and servants. Estimates suggest that tens of thousands of African Americans were forced to work in these non-combat roles. There were isolated cases of African Americans fighting for the Confederacy, but these were exceptional and not part of an organized recruitment effort.

In summary, while the Union Army actively recruited African Americans and formed separate regiments for them, the Confederate forces did not formally include African American soldiers in combat roles but employed them in non-combat positions. As such, the Union forces had a considerably larger number of African Americans serving in military capacities during the Civil War.

The highest-ranking African American in the Union Army during the Civil War was Major Martin Delany. Delany was appointed as a major of the 52nd U.S. Colored Infantry, making him the highest-ranking African American field officer in the Union Army. He was also the highest-ranking African American officer to hold a command position during the war. Delany played an important role in advocating for the recruitment and equal treatment of African American soldiers in the Union forces. However, it is worth noting that there were African American individuals who held higher ranks in non-combat roles, such as surgeons and chaplains.

There were several key Confederate figures who played prominent roles during the American Civil War. Here are some of the most notable ones:

1. Jefferson Davis: Davis served as the President of the Confederate States of America from 1861 until 1865. Before the war, he had been a Mississippi

Senator and Secretary of War under President Franklin Pierce.

2. Robert E. Lee: Lee is widely regarded as one of the most skilled generals of the Confederate Army. He served as the commander of the Army of Northern Virginia and led numerous successful campaigns, earning the respect of both Confederate and Union troops. Lee surrendered to Union General Ulysses S. Grant at Appomattox Court House in 1865.

3. Thomas "Stonewall" Jackson: Jackson was another distinguished Confederate general who served under Robert E. Lee. His military abilities and tactical innovations made him a highly respected commander. He tragically died in 1863 after being accidentally shot by his own troops during the Battle of Chancellorsville.

4. J.E.B. Stuart: Stuart was a renowned Confederate cavalry officer who became known for his daring raids and reconnaissance missions. He played a significant role in many battles, including the First Battle of Bull Run and the Battle of Gettysburg.

5. Nathan Bedford Forrest: Forrest was a controversial figure, known for his relentless tactics and exceptional cavalry leadership. He was involved in numerous successful raids and battles, but his legacy is also marred by his association with the Ku Klux Klan after the war.

6. Alexander Stephens: Stephens served as the Vice President of the Confederate States of America. He was an influential figure in the Confederate government and was tasked with promoting and defending the Confederate cause.

These are just a few of the key figures in the Confederate leadership. There were many other important generals, politicians, and military leaders who played significant roles in the Confederacy during the American Civil War.

During the American Civil War, Ulysses S. Grant, who served as a Union general, had several notable Confederate rivals. The most renowned among them were:

1. Robert E. Lee: Lee, the commander of the Confederate Army of Northern Virginia, was arguably Grant's most significant rival. The two famously clashed in several major battles, including the Overland Campaign and the siege of Petersburg. Their rivalry ultimately culminated in Lee's surrender to Grant at Appomattox Court House in April 1865.

2. Joseph E. Johnston: Johnston was one of the Confederate Army's top generals and commanded the Army of Tennessee. He competed with Grant during the Vicksburg Campaign and later faced him again during the Atlanta Campaign. Despite his tactical skills, Johnston was eventually replaced by General John Bell Hood, who adopted a more aggressive approach.

3. John Bell Hood: Hood succeeded Johnston as the commander of the Confederate Army of Tennessee. He engaged in a series of battles against Grant's forces during the Atlanta Campaign, culminating in the Battle of Franklin and the Battle of Nashville, both of which ended in Confederate defeats.

4. Jubal Early: Early was a Confederate general who led the Army of Northern Virginia on expeditions into the Shenandoah Valley. He attempted to divert Union forces away from Grant's operations in 1864 by threatening Washington, D.C. Early's campaigns, though initially successful, were ultimately repelled by Union forces led by Grant and other Union generals.

While these were Grant's most notable Confederate rivals, it's important to note that Grant faced numerous other Confederate generals and commanders throughout the war. These rivalries contributed to the complex dynamics and strategies of the Civil War, ultimately shaping its outcome.

Ulysses S. Grant's experiences in the military played a crucial role in shaping his path to the presidency. Here are some key ways his military career influenced his rise to power:

1. Leadership and Victories: Grant's military career showcased his ability to lead effectively and achieve significant victories. During the American

Civil War, Grant rose through the ranks and became known for his successful campaigns, such as the capture of Fort Henry, Fort Donelson, and Vicksburg. These achievements established him as a capable military strategist and garnered him a reputation for decisive action.

2. Recognition and Prominence: Grant's military successes gained him recognition and put him in the national spotlight. As news of his victories spread, he became a widely celebrated figure and enjoyed significant popularity among the Northern population. Grant's growing fame and positive public image were important factors in his subsequent political career.

3. Union Hero: Grant's role as a Union hero endeared him to the Republican Party, particularly as the party sought a strong and admired leader to guide the nation through the aftermath of the Civil War. The Republican Party recognized Grant's military achievements as an asset and nominated him as their presidential candidate in 1868.

4. Association with Lincoln: Grant had a close working relationship with President Abraham Lincoln during the Civil War, and the two developed a mutual respect and trust. Grant's association with Lincoln, who was a revered figure within the Republican Party, further enhanced his political standing. After Lincoln's assassination in 1865, Grant's position as a military leader and connection to the late president helped his political ambitions.

5. Reconstruction Efforts: As General-in-Chief of the U.S. Army after the Civil War, Grant was responsible for overseeing Reconstruction efforts in the South. His experiences dealing with the challenges of reintegrating the former Confederate states into the Union and addressing civil rights issues gave him firsthand exposure to the complexities of post-war governance. This experience provided Grant with insights into the political landscape and the issues facing the nation.

Overall, Grant's military career established him as a respected and admired figure, leading to his nomination as the Republican presidential candidate and eventually his election as the 18th President of the United States in 1868. His experiences in the military allowed him to leverage his leadership abilities, recognition, and association with Lincoln to attain the highest office in the country.

6
BATTLES FOUGHT

There were many battles fought during the American Civil War. While it is difficult to provide an exact number, estimates suggest that there were over 10,000 engagements, including major battles, sieges, skirmishes, and minor conflicts. Some of the most well-known battles include the Battle of Gettysburg, the Battle of Antietam, the Battle of Bull Run (First and Second), the Battle of Shiloh, and the Battle of Vicksburg, among others. These battles played a crucial role in shaping the outcome of the war and the trajectory of American history.

The Battle of Antietam, also known as the Battle of Sharpsburg, was fought on September 17, 1862, near Sharpsburg, Maryland. The battle took place during the American Civil War. It was the first major battle to take place on Northern soil and resulted in the bloodiest single day in American history, with an estimated 22,717 casualties.

The Battle of Antietam was considered a tactical draw, but it was a strategic victory for the Union Army. Both the Union Army, commanded by General George McClellan, and the Confederate Army, led by General Robert E. Lee, suffered heavy casualties. However, Lee's Army of Northern Virginia was forced to retreat across the Potomac River into Virginia, giving the Union control of Maryland. While the battle did not result in the complete defeat of the Confederate forces, it is considered a turning point in the war and gave President Abraham Lincoln the opportunity to issue the Emancipation Proclamation.

The First Battle of Bull Run, also known as the First Battle of Manassas, took place on July 21, 1861, near Manassas, Virginia. It was the first major battle of the American Civil War. The Confederate Army, commanded by General P.G.T. Beauregard, clashed with Union forces under the leadership of General Irvin McDowell. The battle ended in a

Confederate victory, with the Union forces retreating back towards Washington, D.C.

The Second Battle of Bull Run, also known as the Second Battle of Manassas, occurred from August 28 to August 30, 1862, in Prince William County, Virginia. This battle was fought between the Confederate Army, led by General Robert E. Lee, and the Union Army, commanded by General John Pope. The Confederate forces emerged victorious, and the Union army was once again forced to retreat back towards Washington, D.C. The Second Battle of Bull Run was a significant Confederate success and allowed Lee to launch his first invasion of the North during the Civil War.

The Battle of Fredericksburg, fought on December 11-15, 1862, during the American Civil War, resulted in a Confederate victory. The battle took place in and around Fredericksburg, Virginia, with Union forces led by General Ambrose Burnside and Confederate forces commanded by General Robert E. Lee.

The Union forces launched a series of assaults against well-entrenched Confederate positions on Marye's Heights, but they were repulsed with heavy losses. The Confederate defense, bolstered by artillery and advantageous positions, proved too formidable for the Union forces to overcome. The Battle of Fredericksburg resulted in high casualties for the Union and little significant gain.

The Confederate victory at Fredericksburg was a significant setback for the Union Army and further solidified General Lee's reputation as a formidable commander. The battle showcased the challenges faced by attacking forces against strong defensive positions and emphasized the importance of strategic positioning and utilizing advantageous terrain during combat.

The Battle of Chancellorsville, fought from April 30 to May 6, 1863, during the American Civil War, resulted in a Confederate victory. The battle took place around the small town of Chancellorsville, Virginia.

Confederate General Robert E. Lee, leading the Army of Northern Virginia, faced Union General Joseph Hooker and the Army of the Potomac. Despite being outnumbered, Lee devised a daring plan, dividing his forces to launch a surprise flank attack. The Confederate forces, through aggressive and bold tactics, were able to drive back the Union army and force them to retreat.

The Battle of Chancellorsville is considered one of Lee's greatest tactical victories, but it also came at a high cost. Lee's second-in-command, General Thomas "Stonewall" Jackson, was accidentally shot by his own men during the battle and later died of his injuries. Despite this loss, the Confederate victory at Chancellorsville boosted morale among Confederate forces and showcased Lee's tactical prowess.

The Battle of Vicksburg, which took place from May 18 to July 4, 1863, during the American Civil War, resulted in a Union victory. The Union forces, led by Major General Ulysses S. Grant, laid siege to the Confederate-controlled city of Vicksburg, Mississippi. After six weeks of fighting and blockade, General John C. Pemberton, the Confederate commander, surrendered the garrison on July 4, 1863. The Union victory at Vicksburg was a significant turning point in the war, as it granted the Union control of the Mississippi River and divided the Confederacy in two.

The Battle of Chickamauga, fought on September 19-20, 1863, during the American Civil War, resulted in a Confederate victory. The battle took place near Chickamauga Creek, Georgia, and involved the Union Army of the Cumberland, commanded by General William Rosecrans, and the Confederate Army of Tennessee, led by General Braxton Bragg.

During the battle, the Union forces suffered significant casualties and were eventually forced to retreat. The Confederate forces, although also experiencing heavy losses, were able to hold the field and claim victory. The Battle of Chickamauga was one of the bloodiest battles of the war, with over 34,000 casualties. While the Confederates emerged as the winners, they were unable to fully capitalize on their victory and were subsequently besieged at Chattanooga, Tennessee, by the Union forces.

The Battle of Franklin, fought on November 30, 1864, during the American Civil War, resulted in a Union victory. The battle took place near Franklin, Tennessee.

The Confederate forces, commanded by General John Bell Hood, launched a series of frontal assaults against the fortified positions of the Union Army of the Ohio, led by General John M. Schofield. Despite the determined efforts of the Confederate troops, they were unable to break through the Union defenses, and their assault resulted in heavy casualties.

The Union victory at the Battle of Franklin halted General Hood's advance towards Nashville and dealt a significant blow to the Confederate Army. The Confederates suffered over 6,000 casualties, including several high-ranking officers. Although the battle did not spell the end of the Confederate Army, it marked a significant setback and further weakened their forces.

The Battle of Hampton Roads, which took place during the American Civil War, was a stalemate between the ironclad warships USS Monitor and CSS Virginia (formerly USS Merrimack). As neither side gained a clear victory, it is often considered a draw.

The Battle of Gettysburg is widely regarded as the bloodiest battle of the American Civil War. It took place from July 1 to July 3, 1863, near the town of Gettysburg, Pennsylvania. The battle resulted in an estimated 51,000 casualties, including killed, wounded, and missing soldiers on both sides. This battle was a turning point in the war and is often considered a Union victory.

Abraham Lincoln felt the need to address the Battle of Gettysburg because it was a pivotal moment in the American Civil War and carried significant moral, political, and symbolic importance. The battle, which took place from July 1 to July 3, 1863, resulted in a Union victory but came at a great cost with heavy casualties on both sides.

Lincoln delivered his Gettysburg Address on November 19, 1863, during the dedication ceremony for the Soldiers' National Cemetery at Gettysburg. His speech aimed to honor the fallen soldiers, honor the sacrifice they made, and emphasize the importance of preserving the Union and upholding the principles of democracy.

At the time, the nation was deeply divided, and the conflict over slavery and states' rights had reached a critical point. The Battle of Gettysburg was seen as a turning point in the war, and Lincoln sought to focus the nation's attention on the higher purpose for which the soldiers had fought and died. He aimed to inspire unity, reaffirm the values of equality and democracy, and express his commitment to the preservation of the Union.

Lincoln's address emphasized the idea that the nation was "dedicated to the proposition that all men are created equal" and that the sacrifices made by the soldiers on the battlefield should not be in vain. It became one of the most famous speeches in American history, encapsulating the spirit of the Civil War and reminding the nation of its shared purpose and aspirations.

In summary, Lincoln felt the need to address Gettysburg to bring a sense of meaning and purpose to the tremendous loss of life in the battle and to reinforce the ideals for which the soldiers fought and died.

The Gettysburg Address is one of the most famous speeches in American history, delivered by President Abraham Lincoln on November 19, 1863. It was given at the dedication ceremony for the Soldiers' National Cemetery in Gettysburg, Pennsylvania, where one of the bloodiest battles of the American Civil War had taken place earlier in the year.

In his address, Lincoln reflected on the significance of the battle and the sacrifice of those who fought and died there. The speech lasted only a few minutes but is considered a powerful and concise articulation of the principles of democracy, equality, and the preservation of the Union.

Lincoln emphasized the importance of preserving the Union and the idea that the United States is "a government of the people, by the people, for the people," as stated in the opening of the speech.

Lincoln spoke about the principles of equality and the fairness that all individuals should be afforded. He invoked the ideals of the Declaration of Independence, declaring that the nation was dedicated to the proposition that "all men are created equal."

The speech acknowledged the immense sacrifice made by the soldiers who fought at Gettysburg, emphasizing that their contributions and sacrifices would not be in vain. Lincoln called upon the living to ensure "that these dead shall not have died in vain" by continuing the struggle for freedom and democracy.

The Gettysburg Address quickly became regarded as a masterpiece of American oratory. It symbolizes the enduring principles of democracy and the importance of national unity, even in the midst of the devastating and divisive Civil War.

7
MILITARY STRATEGY

President Abraham Lincoln and General Ulysses S. Grant communicated regularly during the Civil War. Lincoln appointed Grant as commander of all Union armies in 1864, and they maintained an active correspondence discussing military strategy and planning. They often exchanged letters and telegrams discussing the progress of the war and coordinating their efforts.

Ulysses S. Grant was a prominent figure in American history, particularly during the Civil War. He served as the commander of the Union Army and later became the 18th President of the United States.

Grant's military career began in the Mexican-American War, where he displayed bravery and leadership skills. However, he gained much of his fame during the Civil War, where he played a crucial role in securing Union victory. Here's how he was pivotal to the war:

1. Victories in the Western Theater: Grant achieved notable successes in the Western Theater of the war. He won the first major Union victories at the Battle of Fort Donelson and the Battle of Shiloh in 1862. These victories helped secure the Union's control over strategic areas in Tennessee and Mississippi.

2. Vicksburg Campaign: Grant's most significant military achievement was the capture of Vicksburg, Mississippi. After a long and intense campaign, his forces successfully besieged the Confederate stronghold, securing control of the Mississippi River. This victory effectively split the Confederacy in two and gave the Union control over a vital supply route.

3. Overland Campaign and Siege of Petersburg: In 1864, Grant took command of all Union forces, leading the Army of the Potomac in the Eastern Theater. His strategy during the Overland

Campaign aimed to grind down and exhaust Confederate forces, even in the face of heavy casualties. Although the campaign resulted in heavy losses for both sides, Grant persisted, effectively keeping Confederate General Robert E. Lee's army under constant pressure. Grant then conducted a lengthy siege on Petersburg, leading to the eventual fall of the city in April 1865.

4. Appomattox Court House: The final and perhaps most pivotal moment of Grant's military career came when he accepted the surrender of Robert E. Lee at Appomattox Court House in April 1865. This event marked the beginning of the end of the Civil War, as it signaled the collapse of the Confederacy and led to the eventual reunification of the United States.

Grant's success as a military strategist and his determination to press forward, despite heavy casualties, played a crucial role in securing Union victory and bringing an end to the Civil War. His achievements on the battlefield ultimately contributed to the preservation of the United States as a single, indivisible nation.

In terms of overall strength, the Union had a stronger cavalry during the American Civil War. The Union had more resources, a larger population, and greater industrial capacity, which allowed them to maintain larger and better-equipped cavalry forces. The Union cavalry often had access to superior weaponry, more horses, and better equipment compared to the Confederate cavalry.

However, it's important to note that the Confederate cavalry had some exceptional leaders, such as General J.E.B. Stuart, who employed innovative strategies and tactics to make the most of their limited resources. Confederate cavalry units played a significant role in the early years of the war, providing valuable intelligence, disrupting Union supply lines, and supporting infantry units in battle.

As the war progressed, the Union cavalry grew stronger and more effective, playing a crucial role in countering Confederate raids and providing mobility on the battlefield. The Union cavalry's ability to project power and conduct effective campaigns, such as those led by General Philip

Sheridan, further demonstrated their strength in comparison to the Confederate cavalry.

The railroad system played a crucial role in the American Civil War and was instrumental in the movement of troops, supplies, and communication. The railroad system provided a faster and more efficient means of transporting troops and essential supplies across vast distances. Troop movements could be expedited, allowing armies to be deployed quickly to key strategic locations. This was especially critical in a war characterized by large-scale mobilization and the need to rapidly reinforce positions. Controlling key railroad lines became vital in determining the outcome of battles and campaigns. Railroads enabled armies to transport supplies and troops to the front lines, ensuring they were well equipped and reinforced. Control over strategic rail lines could severely hamper an enemy's ability to move effectively, essentially confining them to specific areas.

The railroad system facilitated the movement of supplies from manufacturers and agricultural regions to the front lines. This was crucial for both the Union and the Confederacy, as maintaining a steady flow of provisions was essential for sustaining their respective war efforts. The railroads enabled the efficient distribution of food, ammunition, weapons, medical supplies, and other necessities to troops in the field.

Railroads improved communication and allowed for more rapid dissemination of information. Telegraph lines were often laid alongside rail lines, enabling military commanders and government officials to communicate more efficiently. This enhanced the coordination of military operations and helped maintain centralized command and control.

The financial and industrial strength of the Northern states allowed the Union to leverage its extensive railroad network, which provided a significant advantage during the war. The railroads facilitated the movement of raw materials and goods, fueling industrial production and ensuring a steady flow of resources to support the Union war effort.

It is worth noting that while railroads were valuable assets, they also had their limitations. Rail lines could be easily damaged or sabotaged by opposing forces, disrupting transportation and logistics. Additionally, maintaining and repairing the railways required extensive resources and manpower, adding to the logistical challenges faced by both sides during the war.

Overall, the railroad system proved highly useful during the American Civil War. It revolutionized military logistics, enabling faster mobilization, improved supply chains, and enhanced communication. The railroads played a significant role in shaping the outcome of battles and campaigns, especially when it came to the movement of troops and supplies. During the American Civil War, there were several significant railroad incidents that occurred. Here are a few notable examples:

1. The Great Train Chase (1862): During the Union's pursuit of the Confederate General John Hunt Morgan, a group of Union soldiers led by James J. Andrews stole a locomotive named "The General"

in Georgia. They attempted to sabotage Confederate supply lines by destroying railroads and disrupting communication. However, they were eventually captured, and some were executed as spies.

2. The Baltimore Plot (1861): There were rumors of a plot to assassinate President Abraham Lincoln while he was traveling by train through Baltimore to his inauguration. In response, a secret night-time route was planned to avoid possible harm. Lincoln successfully arrived in Washington, D.C., but the incident highlighted the tension and danger surrounding rail travel during the war.

3. The Great Locomotive Chase (1862): Union raiders, led by James J. Andrews, stole another locomotive called "The General" in Georgia with the intention of causing havoc and cutting off Confederate supply lines. They conducted a daring chase and sabotage mission. Although they eventually faced capture and execution, their actions disrupted Confederate transportation temporarily.

4. Raid on Ohio & Mississippi Railroad (1862): Confederate General John Hunt Morgan led a raid on the Ohio & Mississippi Railroad, targeting bridges and tracks to interrupt Union supply lines. The raid caused considerable damage, disrupted transportation, and tied down Union forces as they pursued Morgan and his cavalry raiders.

These incidents demonstrate the strategic importance of railroads during the Civil War and how both sides sought to exploit or disrupt them to gain an advantage in the conflict.

The Navy played a significant role during the Civil War and provided several key contributions that were instrumental to the Union's ultimate victory.

One of the Navy's primary tasks was to establish and maintain a blockade of Confederate ports along the Atlantic and Gulf coasts. This blockade limited the South's ability to import and export goods, severely hampering their economy and war efforts. The blockade, although not foolproof, was essential in weakening the

Confederate infrastructure and isolating Southern states from international trade.

The Navy, along with the Army, engaged in riverine warfare on major inland waterways, particularly the Mississippi River. By gaining control of these waterways, the Union Navy was able to split the Confederacy in half, effectively preventing the movement of supplies and troops by river. This strategy was exemplified by the successful capture of New Orleans in 1862.

The Navy played a crucial role in supporting amphibious operations and providing naval gunfire support during the war. Naval forces often provided cover fire for Union troops during amphibious assaults, such as the capture of Fort Fisher in North Carolina in 1865. This support helped Union forces secure coastal positions and gain control of strategic points.

The Union Navy developed and utilized riverine ironclad vessels, such as the USS Monitor and the USS Cairo, to combat the threat of Confederate ironclads, like the CSS Virginia (formerly the USS Merrimack). These ironclads proved crucial in challenging Confederate riverine defenses and their control over certain areas.

The Navy played a critical role in gathering intelligence through a network of scouts, spies, and blockade runners. Naval personnel often provided valuable information on Confederate activities, ensuring effective Union planning and operations.

The Navy supported land operations by providing gunboats and transport vessels to ferry troops, supplies, and equipment. They also provided artillery support to land-based operations, bombarding enemy positions and providing troops with crucial firepower.

In any event, the Navy's contributions to the Union's efforts in the Civil War were multifaceted and crucial to the ultimate victory. Their actions in imposing the blockade, engaging in riverine warfare, supporting amphibious operations, providing reconnaissance and intelligence, and offering transport and firepower support were key factors in weakening the Confederacy and assisting the Union military strategies.

8
ARSENAL OF WEAPONS

The American Civil War saw the use of a wide variety of weapons by both the Union and Confederate forces. The most common weapons used by soldiers in the Civil War were rifles and muskets. Percussion cap rifles, such as the Springfield Model 1861 and the Enfield Pattern 1853, were standard issue for infantrymen. These rifles were loaded through the muzzle and fired a single projectile.

Both sides used cannons and artillery for long-range and siege warfare. Smoothbore cannons, like the Napoleon 12-pounder, were widely used. Rifled cannons, such as the Whitworth and Parrott guns, provided improved accuracy and range. Artillery played a crucial role in major battles, with each side employing various sizes and types of cannons.

Cavalry troops commonly used carbines, which were shorter and lighter rifles, designed for mounted soldiers. The Union cavalry used the Sharps carbine and Spencer carbine, while the Confederate cavalry favored the Kerr and LeMat carbines.

Revolvers were the primary sidearms for officers and cavalry soldiers. Some of the commonly used revolvers were the Colt Army Model 1860 and the Remington New Model Army Revolver. The Colt Single Action Army Revolver, often called the "Peacemaker," became popular after the war.

Melee weapons, such as swords, bayonets, and sabers, were still used during the Civil War, primarily by cavalry troops and some infantry units. Sabers, like the Union Model 1840 and the Confederate Model 1860, were essential for mounted charges.

The Civil War witnessed the introduction of ironclad warships, including the Union USS Monitor and the Confederate CSS Virginia (built from the scuttled USS Merrimack). These ironclads revolutionized naval warfare and were armed with cannons and heavy artillery.

It's important to note that as the war progressed, new and improved weapons were developed and utilized by both sides to gain a tactical advantage. The use of firearms and artillery significantly shaped the nature of warfare during the Civil War.

The American Civil War resulted in a significant loss of life on both sides. While it is difficult to determine an exact number, estimates suggest that between 620,000 and 750,000 soldiers died during the war. This includes both Union and Confederate forces.

It's important to note that these figures include only military deaths and not civilian casualties or deaths caused by diseases and other war-related factors. The war's duration, the scale of the fighting, and the significant casualties in major battles, such as Gettysburg and Antietam, contributed to the high number of deaths.

The Civil War was one of the deadliest conflicts in American history, surpassing the death toll of all other U.S. wars combined until that point and resulting in immense loss and suffering for the nation.

9
MASS MEDIA

During the Civil War, newspapers played a crucial role in disseminating information and shaping public opinion. They reported on various aspects of the war, including military campaigns, political developments, and social and economic impacts. Newspapers provided detailed accounts of major battles and military campaigns. They reported on troop movements, casualties, and strategic developments. The reporting often varied, depending on the publication's affiliation with the Union or Confederate cause. Some newspapers offered more objective coverage, while others were biased towards their respective sides.

Newspapers reported on the political developments and policies related to the war, including the actions of both the Union and Confederate governments. They covered topics such as the Emancipation Proclamation, conscription, political appointments, elections, and negotiations for peace.

Newspapers expressed their viewpoints through editorials and opinion pieces. These articles argued for or against various issues, including the war itself, slavery, secession, and the role of the government. Editorials often sought to shape public opinion and rally support for a particular cause or position.

Newspapers covered the impact of the war on individuals and communities on the home front. They reported on fundraising efforts, relief work, social changes, economic consequences, and the experiences of civilians affected by the conflict.

Newspapers published transcripts or summaries of important speeches and public addresses by political leaders, military officials, and prominent figures. These speeches were disseminated to a broader audience through newspapers and helped shape public discourse.

Newspapers often published letters from soldiers serving in the war, providing firsthand accounts of their experiences and perspectives. These letters provided a glimpse into the lives of soldiers and their thoughts on the war.

Both Union and Confederate newspapers engaged in propaganda and disinformation campaigns to influence public opinion. They sometimes exaggerated victories or downplayed losses to boost morale and support for their respective causes.

It is worth noting that there was a wide range of newspapers during this period, with varying degrees of accuracy and bias. Some were explicitly partisan and represented the viewpoints of specific political factions. Others aimed for more objective reporting, though biases were still present. Understanding the context and affiliations of individual newspapers is essential when analyzing their coverage of the Civil War.

During the Civil War, people obtained newspapers through various means. Here are some common methods and distribution channels for acquiring newspapers during that period:

1. Newsstands: In urban areas, newsstands were a common way to purchase newspapers. These stands were often located in busy public areas or near transportation hubs, allowing people to conveniently purchase the latest newspapers.

2. Subscriptions: Many individuals subscribed to newspapers and received them by mail. Subscriptions were typically prepaid, and newspapers were delivered on a regular schedule. This method ensured that readers received their newspapers directly at their homes or businesses.

3. Street Vendors: Street vendors or newspaper boys would sell newspapers on the streets. They would often shout headlines or distribute handbills to attract attention and encourage sales. These vendors were particularly prevalent in urban areas where foot traffic was high.

4. Military Distribution: Soldiers in both the Union and Confederate armies received newspapers through military distribution channels. Referred to as camp newspapers, these publications were often

distributed to soldiers to keep them informed about the war and boost morale.

The post office played a significant role in delivering newspapers during the Civil War. The post office was responsible for transporting mail and packages, including newspapers, to various destinations across the country. Subscribers who had prepaid for newspaper subscriptions would often receive their newspapers through the mail delivery system. The post office worked to ensure timely and efficient distribution of newspapers to communities and military encampments. However, it is important to note that the postal service during the Civil War faced challenges due to the war's disruptions, such as the Confederate States of America establishing its own postal service. Despite these challenges, the post office remained a crucial means of newspaper distribution during this period.

The United States Postal Service (USPS) continued to deliver letters during the Civil War. The mail service played a crucial role in connecting people and serving as a means of communication during this time of conflict. Despite the challenges posed by the war, the USPS made efforts to maintain its operations as best as possible.

However, the functioning of the postal service was not without difficulties. The war disrupted transportation networks, including railways and roads, making it harder to deliver mail across long distances. In some regions, especially those directly affected by battles or under Confederate control, mail delivery was more sporadic or interrupted altogether.
Nevertheless, the USPS continued to operate and adapt during the war. Postal routes were adjusted, and alternative means of transportation, such as boats or couriers, were utilized when necessary. The importance of the mail service for soldiers on both sides of the conflict led to the establishment of military post offices to ensure that soldiers could communicate with their families and receive letters from home.

Despite the challenges faced by the USPS, postal services remained a vital means of communication for civilians and soldiers during the

Civil War, connecting people across the nation even in the midst of conflict.

Letters were intercepted during the Civil War. Both the Union and Confederate forces engaged in the interception and inspection of mail to gain intelligence or prevent potentially sensitive information from reaching the other side.

The practice of intercepting mail was more common in areas where military operations were taking place or in regions under direct control of the military. Military authorities on both sides established systems for examining and censoring mail as they deemed necessary. Soldiers would often have their letters inspected, and any information that might benefit the enemy, reveal military plans, or contain sensitive information would be removed or censored.

Additionally, both Union and Confederate governments implemented policies to regulate and censor mail to prevent the dissemination of information that could be damaging to their respective causes. The Union established the Office of the Chief of the Secret Service, initially tasked with suppressing counterfeit currency, but later given the responsibility of censoring mail. The Confederacy also employed its own systems for mail censorship.

Intercepted letters were examined by military or government officials, and if deemed appropriate, they were delivered with portions removed or replaced by censorship marks. In some cases, intercepted mail could be completely confiscated or destroyed if it was deemed too risky or of potential value to the enemy.

It is important to note that intercepted mail was not limited to military correspondence but also included civilian letters. The interception of civilian mail aimed to monitor sentiments and opinions, gather intelligence about public sentiment, and prevent the flow of information that could be considered treasonous or harmful to the war effort.

Overall, intercepting and censoring mail were common practices during the Civil War for the purpose of gathering intelligence, maintaining security, and ensuring the continued success of each side's military operations.

10
WAR EFFORT

The American Civil War was a complex conflict with various dynamics.The issue of slavery was at the core of the war, and it shaped the political, economic, and social dynamics of the nation. The debate over the expansion of slavery into new territories and states divided the country.

The war further deepened the divide between states' rights and a centralized federal government. The Southern states believed in the states' right to secede, while the Union sought to preserve the authority of the federal government.

The North was industrializing rapidly, while the South remained predominantly agrarian. This economic difference led to contrasting priorities and perspectives on issues such as tariffs, infrastructure, and labor.

The war witnessed various military strategies and advancements. Initially, there was a belief in quick victories, but it evolved into a long and brutal war of attrition. New tactics and technologies, such as trench warfare, ironclad ships, and rifles, reshaped the battlefield.

The international community closely followed the Civil War, as countries decided whether to support the Union or the Confederacy. The Union managed to prevent official recognition of the Confederacy by foreign powers, but the war had implications for global politics and economies.

As the war progressed, the issue of emancipation became central. President Abraham Lincoln's Emancipation Proclamation and the eventual passage of the Thirteenth Amendment resulted in the abolition of slavery. The war also played a crucial role in the fight for civil rights and the eventual end of racial segregation in the United States.

These dynamics, among others, contributed to the complexities and significance of the American Civil War.

During the American Civil War, the typical meals for soldiers and civilians varied depending on their circumstances and resources. The availability of food was often scarce and inadequate, especially for soldiers on the front lines and civilian populations affected by the war.

Hardtack, also known as "pilot bread" or "hardtack biscuit," was a staple food for soldiers. It was a tough, dry cracker made from flour, water, and sometimes salt. Soldiers would often have to soak the hardtack to make it edible.

Dried and salted meat, such as salt pork or beef, was a common part of soldiers' rations. It could be cooked or boiled to add flavor to the often plain and monotonous meals.

Beans and rice were inexpensive and provided a good source of sustenance. Soldiers and civilians alike often relied on these affordable staples for a basic meal.

Cornmeal was a readily available and versatile ingredient. It could be used to make cornbread, fried into Johnny cakes, or boiled into a mush-like porridge.

Fresh vegetables were often scarce, particularly for soldiers on the move. Canned or pickled vegetables, such as pickled cabbage or sauerkraut, were more common. Dried or canned fruits were also occasionally available.

Coffee became a staple beverage for soldiers, providing a boost of energy and warmth. Soldiers often received coffee as part of their rations. Tea was also consumed, typically made from dried herbs and leaves.

In desperate situations, soldiers and civilians resorted to foraging for wild foods. This could include berries, nuts, wild game, and other edible plants.

It's important to note that the availability and variety of food varied greatly depending on the region, access to supply lines, and the specific circumstances of individuals. The war greatly disrupted agricultural production and transportation, leading to widespread food shortages and hunger. Soldiers and civilians often had to make do with limited resources and adapt to the circumstances to survive.

The cost to outfit a soldier during the Civil War and today can vary significantly due to factors such as inflation, technological advancements, and changes in military equipment and gear. However, it is possible to provide a general comparison.

During the Civil War, the average cost to outfit a Union soldier was around $175 to $225. This would include basic items such as a uniform, shoes, a cap, a knapsack, a blanket, a mess kit, a cartridge box, a bayonet, and a rifle. However, it is important to note that these costs could vary based on factors such as quality, availability, and whether the soldier was a volunteer or conscript.

In contrast, today's outfitting costs for a soldier can be significantly higher. The exact expenses depend on the country, military branch, and position, as well as the specific equipment and gear required. Full military gear and equipment for a modern soldier can range from a few thousand to tens of thousands of dollars. This includes items such as uniforms, body armor, helmets, boots, weapons, communication devices, night vision goggles, and other specialized equipment.

Moreover, modern military technology, such as advanced weapons systems, aircraft, vehicles, and communication systems, greatly increases the overall cost of outfitting a modern soldier compared to the simpler gear used during the Civil War.

Overall, while it is challenging to provide an exact comparison, it is safe to say that outfitting a soldier today is considerably more expensive than it was during the Civil War due to inflation, technological advancements, and the complexity of modern military equipment.

During the Civil War era in the United States, the cost of buying and maintaining a slave included more than just the initial purchase price.

As mentioned earlier, the price of a slave could range from a few hundred dollars to a few thousand dollars, depending on various factors such as location, age, skills, health, and demand.

Slaveowners were responsible for providing food, clothing, and basic necessities to their enslaved individuals. Estimates suggest that this could cost around $10 to $20 per year per person. However, not all slaveowners provided adequate provisions, leading to malnutrition and poor living conditions for many enslaved individuals.

Slaveowners needed to provide housing and shelter for their enslaved individuals. The cost would vary depending on the region and the type of housing. Estimates suggest that this could range from a few dollars to around $50 per year per person. Enslaved individuals often lived in poor conditions and were susceptible to various health issues. Slaveowners were responsible for providing medical care when needed, which could include the cost of physicians, medicines, and treatments, estimated to range from a few dollars to around $50 per year per person. Additional costs could include any necessary equipment or tools for enslaved individuals' work, transportation fees, and fees for slave patrols or other forms of law enforcement.

It is essential to note that these figures can vary widely, and the estimates provided are general averages based on historical records and research. However, it is crucial to recognize that slavery was a deeply unjust system that stripped away the humanity and inherent rights of enslaved individuals, reducing them to mere property. No price or cost can justify or accurately represent the immense human suffering and moral wrongness of slavery.

11 SLAVERY

Slavery was deeply rooted in the Southern United States by the time the Civil War emerged. It had been a significant feature of the region's economy and society for several centuries. Slavery was first introduced to the North American colonies by the European settlers in the early 17th century. Over time, it evolved into a system of forced labor primarily based on racial chattel slavery, where individuals were considered property and could be bought, sold, and owned for life.

In the Southern colonies, the agricultural economy, particularly the cultivation of labor-intensive crops such as tobacco, rice, and indigo, led to an increasing demand for enslaved labor. Slavery was seen as essential for economic prosperity, and it became deeply ingrained in the fabric of Southern society. Slavery provided the foundation for the region's social hierarchy, as well as its political and economic power.

The Southern states heavily relied on enslaved labor to work on plantations and in other industries, such as mining and construction. Enslaved individuals were considered valuable property and were bought and sold as commodities. It was a brutal and dehumanizing system, with enslaved people being subjected to physical and psychological abuse, separated from their families, and denied basic human rights.

By the mid-19th century, as the United States expanded westward, the institution of slavery profoundly influenced the political landscape of the country. The South, with its agrarian economy dependent on slave labor, became increasingly defensive of its "peculiar institution" and sought to protect and expand it into new territories. This led to growing tensions between the Northern states, where industrialization and a more diverse economy began to erode the reliance on slavery, and the Southern states, where the institution remained deeply entrenched in social, economic, and political life.

The issue of slavery became a central point of contention that ultimately led to the secession of Southern states and the eruption of the Civil War in 1861. Thus, it can be said that slavery was deeply rooted in the Southern United States by the time the Civil War emerged, shaping the region's identity, economy, and society.

The first enslaved Africans are believed to have arrived in the territory that would become the United States in August 1619. A ship known as the White Lion, captained by English privateers, brought around 20 enslaved Africans to Point Comfort in the English colony of Virginia. These individuals were sold into servitude, marking the beginning of the institution of slavery in what would later become the United States. It's important to recognize that slavery was already a deeply entrenched practice in various parts of the world prior to this arrival. The arrival of enslaved Africans in the Virginia colony, however, played a significant role in shaping the future of slavery in America.

Slavery was economically beneficial to the South in the antebellum period, primarily due to the labor-intensive nature of the region's agrarian economy. The Southern states heavily relied on cash crop agriculture, particularly cotton, tobacco, rice, and sugar cane, which required significant amounts of labor to cultivate and harvest.

Enslaved labor provided a cheap and captive workforce for plantation owners and farmers. This allowed them to maximize their profits by reducing labor costs and increasing agricultural productivity. The profitability of the Southern slave economy was fueled by the high demand for cotton, both domestically and internationally, which led to expanding plantation systems and increased reliance on enslaved labor.

Slavery also played a critical role in building the Southern economy and generating wealth. Enslaved individuals were considered property and could be bought, sold, and mortgaged. The ownership of slaves allowed wealthy landowners to amass large plantations and accumulate considerable wealth and social status.

Additionally, the cotton industry, fueled by enslaved labor, played a significant role in the growth of the Northern textile industry. Cotton production in the South provided raw materials for Northern mills, contributing to the industrialization and economic growth of the North.

During the Civil War era in the United States, owning slaves was predominantly a practice among the wealthy and upper-class individuals. Slave ownership was generally associated with the plantation owners of the Southern states, who formed the elite class known as the planter class. These individuals had the financial means and resources to afford and maintain slaves.

On the other hand, the majority of the working-class population, including poor farmers, laborers, and small-scale landowners, did not have the financial means to purchase and maintain slaves. Slave ownership was largely limited to the elite class due to the significant expense involved in purchasing and supporting enslaved individuals.

However, it is worth noting that not all members of the planter class could afford slaves either. Slave ownership was concentrated among the wealthiest landowners, while many smaller landowners aspired to own slaves but could not afford to do so.

It is important to recognize that while the ability to own slaves was primarily limited to the wealthy, this does not absolve them from the moral wrongdoing and inhumanity of participating in a system that treated human beings as property. Slavery was a deeply unjust and dehumanizing institution that perpetuated inequality and systemic oppression.

However, it's crucial to note that the economic benefits derived from slavery were concentrated in the hands of a small elite minority of white plantation owners, while the majority of Southerners did not own slaves and did not benefit directly from the institution.

Furthermore, while the economic benefits of slavery were undeniable in the short term, there were long-term negative consequences. The reliance on slavery hindered the region's economic and industrial development as investment in other sectors, such as manufacturing and infrastructure, was comparatively limited. Ultimately, the system of slavery contributed to deep societal divisions, ethical debates, and, eventually, the outbreak of the American Civil War.

During the Civil War era in the United States, slave ownership was primarily concentrated among a specific demographic group known as the planter class. This class consisted of large-scale plantation owners who held significant wealth and power within the Southern states. However, it is important to note that not all members of this class owned slaves, and there were exceptions within other groups as well.

Wealthy plantation owners with large agricultural holdings, particularly those engaged in cash crops such as cotton, tobacco, or rice, were more likely to own slaves. These individuals formed the economic and social elite of the Southern states and often inherited both land and enslaved labor.

Beyond plantation owners, other large landowners who did not necessarily engage in intensive agriculture, such as prominent families, wealthy businessmen, and sometimes even small-scale farmers, may have owned enslaved individuals as a sign of status and wealth. However, the scale of ownership was generally smaller compared to the planter class.

Many politicians and members of the Southern elite owned slaves. Notable figures like Thomas Jefferson and James Madison were slaveholders, and slave ownership was considered a mark of social and economic standing among the Southern elite.

While the majority of slave ownership was concentrated in rural areas, some individuals in cities and towns owned enslaved individuals as domestic servants, skilled workers, or for other purposes. These urban slave owners were often part of the middle or upper-middle class.

It is important to emphasize that not all Americans in the South owned slaves. Many categories of the population did not own slaves, including the majority of the white population who were small farmers, laborers, or craftsmen. Factors such as economic means, social status, and access to land played a significant role in determining the likelihood of slave ownership within specific groups.

In the context of the Civil War era in the United States, immigrants were not prohibited from owning slaves solely based on their immigration status. However, there were certain restrictions and variations in slave ownership among immigrants, depending on the state they resided in and their socioeconomic status.

During that period, each state had its own laws regarding slavery, and some states had restrictions on the rights of immigrants to own slaves. For example, in some Southern states, individuals had to be citizens of the United States or hold a certain residency status to own slaves. Immigrants without citizenship or legal residency may have faced barriers to slave ownership in those states.

Additionally, the likelihood of immigrants owning slaves varied depending on their socioeconomic status. Immigrants who were wealthy or part of the upper class, such as merchants, businessmen, or professionals, were more likely to own slaves, particularly if they resided in areas with significant slave ownership, such as the Southern states. However, the majority of immigrants during this time were not part of this privileged class and did not own slaves.

It's also crucial to note that the majority of immigrants during the Civil War era were not from Western or Northern Europe but mainly from countries in Southern and Eastern Europe. Immigrants from these regions were generally not as wealthy or socially connected as those from Western or Northern Europe, making it less likely for them to own slaves. However, it is important to remember that individual circumstances and factors could influence slave ownership among immigrants, and there may be exceptional cases.

Slaves did run to Native Americans for help during the time of slavery in the United States, including during the Civil War. Native American tribes were often seen as potential allies and places of refuge by escaped slaves seeking freedom.

Native American tribes and communities offered assistance to escaped slaves for various reasons. Some tribes, such as the Seminole in Florida and the Choctaw and Chickasaw in the Indian Territory, had pre-existing relationships with enslaved African Americans. These tribes had a history of intermarriage, shared land, and cultural exchange. As a result, they were more likely to offer refuge and assistance to escaped slaves.

The Seminole tribe, for instance, had a long history of harboring escaped slaves. They welcomed escaped slaves into their communities, where they were granted freedom and protection, often integrating and intermarrying with the tribe. Seminole communities in Florida became a significant destination for escaped slaves during the antebellum period.

During the Civil War, the presence of Union forces in the Indian Territory provided further opportunities for escape. Slaves from surrounding Confederate states, such as Arkansas and Texas, sought refuge with Native American tribes that were aligned with the Union cause. These tribes sometimes provided assistance to escaped slaves, including shelter, food, and protection.

It is important to note that not all Native American tribes or individuals supported or aided escaped slaves. The attitudes and actions toward escaped slaves varied among tribes, depending on tribal customs, cultural beliefs, and relationships with slaveholding Southern states. Some tribes, particularly those with close ties to Southern society, were not as welcoming and would return escaped slaves to their owners.

Overall, while Native Americans were not uniformly supportive, many tribes and individuals provided assistance and refuge to escaped slaves seeking freedom during the time of slavery and the Civil War. The relationships that developed between Native Americans and escaped slaves highlight a historical context of shared struggles against oppression.

Native Americans played various roles during the Civil War, depending on their tribal affiliations and individual circumstances. It is important to note that Native American tribes were diverse and had unique relationships with both the Union and Confederate governments.

Some Native American tribes, such as the Five Civilized Tribes (Cherokee, Choctaw, Chickasaw, Creek, and Seminole) in the Indian Territory (present-day Oklahoma), were caught in the middle of the conflict. These tribes had developed complex relationships with the United States, including the establishment of treaties and their own governments. Unfortunately, the Civil War disrupted these relationships, as some tribes were divided in their loyalties, with individuals supporting either the Union or the Confederacy.

Many Native Americans joined the Confederate cause due to their cultural and economic ties with Southern society and the belief that an independent Confederate nation would respect their sovereignty and protect their interests. Notable Native American leaders, such as Stand Watie of the Cherokee Nation and Opothleyahola of the Creek Nation, took opposing sides. Stand Watie led Cherokee, Creek, and Seminole troops under Confederate command, while Opothleyahola, along with his followers, remained loyal to the Union and fought against the Confederacy.

On the Union side, Native Americans also contributed to the war effort. Some individuals enlisted in Union regiments and served alongside white soldiers. The most well-known Union Indian regiment was the 1st Regiment Indian Home Guard, formed by loyal Native Americans in the Indian Territory. They fought to protect their lands, tribes, and support the Union cause.

Throughout the war, Native American communities faced immense challenges. Many found themselves caught in the crossfire, with their lands being subjected to violence, displacement, and destruction. Battles and military operations took place on and near Native American lands, impacting their lives and livelihoods.

In summary, Native Americans played diverse roles during the Civil War. Some tribes were divided in their loyalties, while others sought to protect their interests or support the side they believed would respect their sovereignty. Native Americans faced immense challenges and consequences as the war impacted their lands and way of life. Their experiences during this period were varied and complex, reflecting the diversity of Native American tribes and their relationships with the United States.

Before the Civil War, the per capita slave population in the United States varied by region. The highest concentrations of enslaved individuals were found in the Southern states, particularly in states where large plantation agriculture, such as cotton, thrived.

According to historical data, in 1860, just before the outbreak of the Civil War, the highest per capita slave populations were in states like Mississippi, South Carolina, and Louisiana. In Mississippi, for example, the per capita slave population was approximately 55%, meaning that for every 100 people living in Mississippi, about 55 were enslaved. In South Carolina, the per capita slave population was around 47%, and in Louisiana, it was around 46%.

In some of the other Southern states, the per capita slave population was lower, but still significant. For instance, in Alabama, the per capita slave population was about 40%, in Georgia it was approximately 38%, and in Florida it was around 34%.

In contrast, the per capita slave population was significantly lower in the Northern states, where industrialization and urbanization were more prevalent. The Northern states generally had fewer enslaved individuals, with some states having a very small or negligible slave population.

It's important to note that these figures refer only to the enslaved population and do not reflect the overall population of each state. The per capita slave population gives us an idea of the proportion of enslaved individuals within each state, highlighting the significant presence of slavery in the Southern states before the Civil War.

Quantifying the exact monetary value that slavery contributed to the southern economy is a complex task. While slavery played a crucial role in the accumulation of wealth in the South, it is challenging to assign a precise dollar amount to its economic impact. Additionally, estimating the economic value of an institution rooted in the subjugation and exploitation of human beings raises ethical concerns.

That being said, historians and economists have made attempts to analyze the economic significance of slavery. One widely cited study by economists Claudia Goldin and Robert A. Margo estimated that in 1860, the total economic value of slaves in the United States was approximately $3.5 billion (in 1860 dollars), which exceeded the total value of all the factories, railroads, and productive investments combined at that time.

12
SOCIAL LIFE

During the Civil War era in the United States, there were several prominent families that were known for their significant ownership of enslaved individuals. It is challenging to single out one family as the absolute "most" prominent slave owner, as slave ownership was widespread among the Southern planter class.

That being said, some families gained particular notoriety for their extensive slave ownership and prominent roles in the institution of slavery. One example is the Butler family of South Carolina, which included figures such as Pierce Butler and his grandson, Senator Andrew Butler. The family owned plantations with hundreds of enslaved individuals and played significant roles in shaping Southern politics and defending slavery.

Another notable example is the Custis-Lee family of Virginia, which included George Washington Parke Custis (the step-grandson of George Washington) and Robert E. Lee. The family owned a large number of enslaved individuals on estates such as Arlington House, and Robert E. Lee famously inherited and eventually freed the enslaved individuals who were part of the estate.

These are just two examples, and there were undoubtedly other prominent slave-owning families throughout the Southern states during that time. It is important to remember that the prominence of these families was rooted in their participation in a deeply unjust and oppressive institution. Slavery was a grave violation of human rights, and its prominent families were complicit in perpetuating that injustice.

The family often cited as the most prominent and largest slaveholding family during the Civil War era was the Ball family of South Carolina. The Ball family, specifically their Magnolia Plantation near Charleston, owned an extensive number of enslaved individuals.

The patriarch of the Ball family, Col. William Aiken, was a successful planter and politician. His sons, William Aiken Jr. and John Robinson Aiken, continued their family's legacy as prominent slaveholders. The Aiken family's wealth and power were derived from their extensive holdings of both land and enslaved labor.

It is estimated that at its peak, the Ball family owned around a thousand enslaved individuals, making them one of the most prominent slaveholding families in the South. This level of ownership was not typical, as most slaveholders owned smaller numbers of enslaved individuals.

While the Ball family is often highlighted for their vast ownership of enslaved people, it is essential to remember that the institution of slavery itself was a grave violation of human rights and fundamentally unjust. The size and prominence of the Ball family's slave ownership do not diminish the inherent oppression and suffering endured by the enslaved individuals who were held in bondage on their plantations.

Slavery contributed to economic growth in several ways. It provided an abundant labor force and allowed for the production of cash crops, such as cotton, on large plantations. These crops were exported and generated substantial profits for plantation owners and traders. The cotton industry, in particular, drove the southern economy and played a significant role in U.S. trade and industrialization.

However, it is essential to recognize that measuring the economic impact of slavery fails to account for the immense human suffering, the violation of basic human rights, and the long-term detrimental effects on society. Slavery was a deeply immoral and oppressive system, and its legacy continues to shape social and economic disparities in the United States today.

During the Civil War, slaves often found ways to help each other despite the challenges and risks they faced. Slaves often formed close-knit communities and developed networks of communication to share important information about events, news, and opportunities for escape. This communication could take various forms, including secret meetings, covert conversations, or the use of coded messages or symbols. Sharing information enhanced their collective knowledge and helped them make informed decisions about their circumstances.

Slaves who had successfully escaped bondage often served as guides, providing information and assistance to others planning their escape. They shared knowledge about escape routes, safe houses, and other essential resources, helping their fellow slaves navigate the treacherous journey to freedom.

Within plantation communities, slaves offered support and protection to each other. This could involve looking out for one another, sharing resources and knowledge, and providing emotional comfort during difficult times. By forming bonds and relying on each other, they created a sense of community and solidarity.

Slaves engaged in various forms of resistance and collective action, both individual and organized, to improve their circumstances. This could include acts of sabotage, work slowdowns, the theft of food or supplies, feigning illness to avoid work, or even staging rebellions. By joining together and acting collectively, slaves showed solidarity and resistance against their oppressors.

Slaves maintained and shared their cultural traditions, including music, storytelling, and spiritual practices. These traditions provided comfort, inspiration, and a sense of identity and unity during difficult times. By gathering in secret or finding ways to continue their cultural practices, slaves were able to maintain a sense of community and preserve their heritage.

It's important to note that these acts of support and resistance were often carried out covertly, with great risks and consequences if discovered by slave owners. Despite these challenges, slaves demonstrated resilience and an unwavering commitment to helping one another during the Civil War.

During the Civil War, several African Americans played influential roles in various aspects of the conflict. Here are a few notable individuals:

1. Frederick Douglass: Frederick Douglass was a prominent abolitionist and social reformer. During the Civil War, he served as a consultant to President Abraham Lincoln and advocated for the recruitment of African American soldiers in the Union army. He also played a key role in advancing the cause of emancipation and equal rights for African Americans.

2. Harriet Tubman: Harriet Tubman was an abolitionist and conductor of the Underground Railroad, a network of safe houses and routes that helped enslaved individuals escape to free states or Canada. During the Civil War, Tubman worked as a nurse, cook, and spy for the Union army. She also assisted in leading a successful raid that freed over 700 enslaved people on a plantation in South Carolina.

3. Robert Smalls: Robert Smalls was an enslaved man who escaped slavery by commandeering a Confederate transport ship, the CSS Planter, and delivering it to Union forces. He then served as a pilot for the Union navy and went on to become one of the first African Americans elected to the U.S. House of Representatives.

4. Sojourner Truth: Sojourner Truth was an abolitionist and women's rights advocate. Although not directly involved in the Civil War, she spoke out against slavery and inequality during this period. Her famous speech, "Ain't I a Woman?", delivered at the Women's Rights Convention in Ohio in 1851, highlighted the struggles faced by African American women and their intersecting identities.

5. Martin Delany: Martin Delany was an African American physician, writer, and abolitionist. He advocated for the enlistment of African American soldiers in the Union army and actively recruited black soldiers. Delany also wrote "The Condition, Elevation, Emigration, and Destiny of the Colored

People of the United States," which discussed plans for the establishment of a separate black nation.

These individuals, among many others, made significant contributions to the cause of freedom, equal rights, and the abolition of slavery during the Civil War era.

It is challenging to provide an exact number for how many slaves were fugitives during the Civil War, as accurate records were not consistently kept. However, it is estimated that tens of thousands of slaves sought freedom by escaping from bondage during this period.

The Underground Railroad, a network of secret routes and safe houses, played a significant role in assisting slaves in their escape. Estimates suggest that between 20,000 and 100,000 enslaved individuals successfully escaped to the North and Canada with the help of the Underground Railroad between the late 18th century and the Civil War.

Additionally, as the war progressed, the Union Army's presence in the Southern states provided an opportunity for slaves to seek liberation. The Union Army often offered protection and assistance to escaped slaves, and these individuals became known as "contrabands of war." It is estimated that thousands of contrabands escaped during the war, seeking refuge behind Union lines, and later many of them enlisted in the Union Army.

While the exact number of fugitive slaves during the Civil War cannot be determined, it is clear that a significant number of enslaved individuals seized opportunities to escape bondage and seek freedom throughout this period.

Harriet Tubman played a crucial role in assisting slaves during the Civil War as a conductor for the Underground Railroad. Harriet Tubman made multiple dangerous trips back to Southern states to guide enslaved individuals to freedom. She would travel at night, using various disguises and secret routes to avoid detection. Tubman's deep knowledge of the region and her resourcefulness were invaluable in ensuring successful escapes.

As a conductor, Tubman offered guidance and support to those seeking freedom. She would meet with individuals and groups of escaped slaves, provide them with instructions on how to navigate their journey, and offer words of encouragement and reassurance.

Tubman utilized her extensive network of contacts along the Underground Railroad to facilitate escapes. She would connect escaped slaves with safe houses, sympathetic individuals, and other conductors who could assist them on their journey to freedom.

Tubman would gather and share information about slaveholders, patrols, and potential dangers along the escape routes. The knowledge she acquired through personal experiences and her connections helped enslaved individuals navigate their journey safely.

Tubman would also provide physical care to those making the journey to freedom. This included organizing shelter, procuring food and supplies, and tending to any injuries or illnesses that occurred during the escape.

It's important to note that Tubman's efforts were not limited to the Civil War period, but spanned several years before and after the conflict. The Underground Railroad had been operating since the late 18th century, assisting enslaved individuals in escaping to the North and Canada. Tubman's bravery, determination, and commitment to freedom made her one of the most renowned figures in the Underground Railroad network, and she continued her advocacy and activism long after the Civil War ended.

When slaves entered a free state during the time of legalized slavery in the United States, their status as slaves would typically be challenged. Free states were territories where slavery was either abolished or prohibited, which meant that the legal protections and rights of slaves could be different from those in slave states.

In most cases, slaves who managed to reach a free state could attempt to claim their freedom based on the laws and policies of that state. Depending on the specific state laws, there were a few possible outcomes:

Some free states had laws that automatically freed any slaves who entered their territory. These slaves were granted immediate freedom upon reaching the free state's jurisdiction.

Other free states had laws in place that phased out slavery over a period of time. This meant that slaves who entered these states may not be immediately emancipated but could potentially gain freedom after a certain period.

Slaveholders from slave states could try to reclaim their escaped slaves through legal means in free states. They could file lawsuits and demand the return of their property under the laws known as the Fugitive Slave Acts, which required the return of escaped slaves to their owners.

In some instances, escaped slaves received support from abolitionists, sympathetic individuals, and the Underground Railroad network. They would be assisted in evading capture and sometimes finding employment and protection in free states.

It is crucial to remember that while escaping to a free state offered a chance at freedom, it did not guarantee safety or permanent emancipation. Some slaves were captured and returned to their owners, despite the efforts of abolitionists and individuals assisting the Underground Railroad.

The status and treatment of escaped slaves in free states were influenced by local laws, socio-political circumstances, and individual attitudes towards slavery. The situation for escaped slaves varied across different regions and time periods. The abolition of slavery in the United States did not occur until after the Civil War with the passage of the Thirteenth Amendment in 1865, officially ending the institution of slavery.

The issue of states' rights versus a federal centralized government was a significant factor in the lead-up to the American Civil War. The Southern states, primarily in the Confederate States of America, argued for greater states' rights, while the Northern states advocated for a stronger federal government.

The Confederate states held the belief that the federal government's encroachment on their states' rights, particularly regarding the institution of slavery, threatened their economic and social systems. They argued that individual states had the right to determine their own laws and policies, including the right to retain and expand slavery.

The Southern states believed that their economies and ways of life were dependent on the institution of slavery, and they saw any attempts by the federal government to regulate or restrict slavery as an infringement on their states' rights. They argued that the Constitution supported the concept of state sovereignty, and that each state had the authority to secede from the Union if it felt its rights were being violated.

Conversely, the Northern states and the federal government believed in a more centralized system, with the federal government wielding greater authority. They argued that the federal government had the power to regulate and control issues that affected the entire nation, such as the expansion of slavery into new territories.

The conflict between states' rights and a centralized government came to a head when several Southern states seceded from the Union and formed the Confederate States of America in 1861. The Civil War was fought, in part, over the question of whether states had the right to secede and maintain their own laws, or whether the federal government had the authority to maintain the Union and enforce federal laws throughout the country.

Ultimately, the outcome of the Civil War solidified the power and authority of the federal government over the states. The Union victory reinforced the idea that the federal government's authority in certain matters, such as issues pertaining to slavery, would supersede state laws and rights.

The abolitionist states played a crucial role in supporting the Union during the American Civil War for several reasons:

1. Recruitment of Soldiers: Abolitionist states such as Massachusetts, Vermont, and New York provided a significant number of volunteers to the Union Army. Many residents of these states strongly opposed slavery and believed in the preservation of the Union. Their commitment to the cause led to large numbers of men enlisting, bolstering the Union's military strength.

2. Financial Support: Abolitionist states also provided substantial financial support to the Union war effort. Through taxes, bonds, and other contributions, citizens and businesses from these states helped finance the war, providing the Union with the necessary resources for procuring supplies, paying troops, and sustaining military operations.

3. Political Influence: The anti-slavery sentiments prevalent in abolitionist states had a significant impact on national politics. The abolitionist movement shaped public opinion and influenced

political decisions, pushing the Union government to take a stronger stance against slavery and enforce policies that aimed to end it. Abolitionist politicians, such as Massachusetts Senator Charles Sumner and President Abraham Lincoln, played key roles in formulating and implementing policies that ultimately led to the Emancipation Proclamation and the passage of the Thirteenth Amendment to the United States Constitution, which abolished slavery.

4. Propaganda Efforts: Abolitionist states actively supported the Union cause through propaganda efforts. Abolitionist newspapers and writers, such as Harriet Beecher Stowe, used their influence and words to highlight the moral imperative of preserving the Union and eliminating slavery. They helped shape public opinion in favor of the Union cause, both domestically and internationally.

5. Underground Railroad: Prior to the Civil War, abolitionist states and their citizens played a significant role in the "Underground Railroad," a network of secret routes and safe houses used by enslaved African Americans to escape to free states

and Canada. This resistance to the institution of slavery demonstrated the commitment of abolitionists to the cause of freedom and equality.

It is important to note that not all states in the Union were abolitionist states, and there were also divisions within these states regarding their stance on slavery. However, the commitment and efforts of abolitionist states and individuals played a vital role in supporting the Union cause and ultimately contributed to the Union's victory in the Civil War.

13 ECONOMICS AND FINANCE

The Industrial Revolution, which commenced in the late 18th century and accelerated throughout the 19th century, brought several significant challenges to the Civil War in the United States. The Industrial Revolution led to a rapid industrialization and urbanization in the Northern states, resulting in a thriving manufacturing economy. In contrast, the Southern states, with their agricultural-based economy reliant on slave labor, did not undergo the same level of industrial transformation. This economic disparity exacerbated tensions between the North and South, contributing to the divide that eventually led to the Civil War.

The increasing industrialization in the North created a burgeoning wage labor system, attracting immigrants and rural Americans seeking employment in factories and industries. This labor system sharply contrasted with the Southern reliance on slavery. The question of whether the expansion of slavery into new territories would disrupt this emerging labor system and economy was a prominent issue that fueled tensions during the Civil War.

The Industrial Revolution brought advancements in transportation, communication, and manufacturing technologies. Railroads and telegraph lines, for example, improved the speed and efficiency of communication and transportation of troops and supplies during the war. The introduction of mass-produced weapons and artillery also transformed the nature of warfare. Both sides had to adapt to these technological changes, leading to new strategies and tactics on the battlefield.

The Industrial Revolution increased the demand for resources such as coal, iron, and other raw materials needed for industrial production. The

North had greater access to these resources, leading to an advantage in terms of industrial capacity and military production. This resource disparity influenced the war effort, as the South struggled to match the North's industrial output.

The Industrial Revolution accelerated globalization, connecting nations through trade and economic ties. During the Civil War, countries like Britain and France, major industrial powers in Europe, were tempted by the South's cotton exports. The South hoped that this reliance on cotton would prompt intervention from these European powers in support of the Confederacy. However, the North's blockade of Southern ports, technological advancements in textile manufacturing, and moral objections to slavery limited international intervention in the conflict.

Overall, the Industrial Revolution brought significant economic, technological, and social changes that affected the course and outcome of the Civil War. The North's industrial advantages, coupled with economic disparities and the slavery issue, played a crucial role in determining the ultimate outcome of the conflict.

The Union was wealthier than the Confederacy during the American Civil War. The Union had a more developed and diversified economy, with a strong industrial base, established financial sector, and advanced transportation infrastructure. These economic advantages provided the Union with greater access to resources, capital, and international markets. In contrast, the Confederacy relied heavily on agriculture and lacked the same level of industrial development and financial resources. The Union's economic strength was a significant factor in their ability to sustain the war effort and ultimately contributed to their victory.

During the American Civil War, the Union had a significant advantage in terms of wealth and economic resources compared to the Confederacy. The economic disparity between the two sides played a crucial role in determining the outcome of the war.

The Union had a far more developed and diversified industrial base compared to the agrarian South. Industries such as manufacturing, textiles, ironworks, and coal mining were concentrated in the Northern states, providing them with a significant advantage in terms of production capabilities and access to necessary resources.

The Northern states possessed a well-established transportation network, including an extensive railroad system and a superior system of roads and canals. This facilitated the movement of troops, supplies, and goods throughout the Union, strengthening their logistical capabilities and overall war effort.

The financial sector in the Northern states was more developed and established. Northern banks had greater access to capital, allowing for easier financing of the war effort through bond issuance and loans. The Union's ability to raise funds and sustain their military operations was thus greater than that of the Confederacy.

While the Confederacy had a predominantly agricultural economy, their agricultural output was hampered by the war. The Union blockade of Confederate ports disrupted the South's ability to export cotton and import necessary goods, significantly impacting their economic resources.

The Union had greater access to international markets and trade. Despite the efforts of the Confederacy to seek recognition and assistance from European powers, the Union's blockade and diplomatic efforts limited the South's ability to engage in legitimate trade and import essential resources.

The Union had a larger population and labor force, providing them with a greater pool of potential soldiers and workers. This allowed the Union to raise larger armies and sustain their military campaigns effectively.

These economic disparities played a significant role in enabling the Union to maintain a prolonged and effective war effort compared to the Confederacy. The Union's economic advantages helped to sustain their military operations and ultimately contributed to their victory in the Civil War

Jefferson Davis, who was the President of the Confederate States of America during the American Civil War, authorized the printing of Confederate money out of necessity. When the Southern states seceded from the Union and formed the Confederacy, they established their own government and sought to create a separate nation. As part of this process, they needed a currency to support their economy and finance their war effort.

Since the Confederacy did not have an established banking system or a central bank at the time, they had to resort to printing their own currency, known as Confederate notes. The Confederate government issued these notes as legal tender, hoping they would be widely accepted within the Confederacy.

However, the printing of Confederate money faced several challenges. One major issue was that the Confederacy lacked the resources and infrastructure to support a currency system adequately. They did not have enough gold or silver reserves to back up the face value of their notes, resulting in severe inflation and loss of confidence in the currency.

Furthermore, the Union's naval blockade of Confederate ports limited the Confederacy's ability to import necessary goods, including traditional printing materials like ink, paper, and printing plates. As a result, the quality of Confederate notes declined over time, with issues of counterfeiting and general lack of trust.

Jefferson Davis and the Confederate government continued to print money to finance their war efforts, but the rampant inflation and lack of value ultimately led to the currency's devaluation and contributed to economic instability within the Confederacy.

The American Civil War was financed by various sources on both sides: the Union and the Confederacy.

Union Financing:

1. Government Bonds: The Union government issued bonds to raise funds for the war effort. These bonds were sold to individuals and institutions, promising repayment with interest after a designated period. The sale of government bonds provided a significant source of funding for the Union.

2. Taxes: The Union imposed various taxes to generate revenue for the war. These included income taxes, property taxes, and excise taxes on goods and services. Taxation played a crucial role in financing the Union war effort.

3. Legal Tender Acts: The Union government passed several Legal Tender Acts, which allowed the issuance of "greenback" currency. These were fiat money, not backed by gold or silver, and were used to meet the financial needs of the war effort.

4. Loans and Bonds from Citizens: Individuals and businesses purchased government-issued war bonds, effectively lending money to the Union

government. This provided additional funding for the war effort.

5. International Loans: The Union government sought loans from European countries, particularly from Britain. Financial institutions and individuals in Europe provided substantial funds to support the Union's war effort.

Confederate Financing:
1. Government Bonds: The Confederate government also issued bonds to generate funds for the war. These bonds were sold to individuals and institutions, promising repayment with interest after a specified period. However, the economic challenges faced by the Confederacy limited the success of bond sales.

2. Taxes and Tariffs: The Confederacy enacted various taxes, including income taxes, property taxes, and import tariffs on goods. These measures aimed to generate revenue for the war effort.

3. Printing Money: The Confederacy relied heavily on printing money to fund its war effort. This led

to severe inflation and a depreciation in the value of Confederate notes.

4. Loans from Citizens: Individuals and businesses in the South loaned money directly to the Confederate government to support the war effort. This type of citizen lending was crucial in financing the Confederate war effort.

5. Seizure of Property: The Confederacy confiscated property and resources from private citizens, including land, crops, and livestock, to support the war effort. This contributed to the financing of the Confederate government.

It is important to note that both sides utilized a combination of strategies to finance the war, but the Union had greater access to financial resources and international support, giving them a significant advantage in funding their war efforts.

During the American Civil War, the Union forces implemented a naval blockade against the Confederacy. It was known as the Union blockade of Confederate ports. The blockade aimed to prevent the South from exporting its agricultural products, specifically cotton, which was a crucial source of revenue for the Confederacy. It also aimed to restrict the importation of necessary supplies and goods by the Confederate states.

The Union blockade was formally established on April 19, 1861, shortly after the Confederate attack on Fort Sumter. Initially, it covered only a few key Southern ports along the Atlantic coast, including Charleston, South Carolina, and Savannah, Georgia. However, as the war progressed, the blockade expanded to cover the entire Southern coastline, including the Gulf of Mexico and the Mississippi River.

The Union Navy employed a variety of vessels to enforce the blockade, including cruisers, gunboats, and ironclads. They aimed to intercept and capture Confederate ships attempting to break through the blockade or engage in illicit trade.

While the blockade did not completely seal off Confederate ports, it had a significant impact on the South's ability to export and import goods.

It hindered the South's economy by limiting their access to international markets, preventing them from exporting cotton for revenue and curtailing their ability to obtain essential supplies. The blockade played a crucial role in crippling the Confederate economy and contributed to the Union's ultimate victory in the war.

Several military strategies played important roles in determining the outcome of the American Civil War. Here are a few key strategies that helped contribute to victory for the Union:

1. Anaconda Plan: The Union's overall strategy, known as the Anaconda Plan, aimed to strangle the Confederacy by utilizing blockades, seizing control of rivers, and splitting the South by capturing key strategic locations. This plan focused on cutting off Confederate supply lines, isolating their forces, and limiting their ability to receive external support.

2. Total War: Union General Ulysses S. Grant adopted a strategy of total war, seeking to destroy the Confederacy's ability to wage war by targeting

valuable resources, infrastructure, communication systems, and civilian support networks. Grant employed aggressive tactics and was relentless in pursuing the Confederates, ultimately wearing them down.

3. Control of Rivers: Control of vital waterways, particularly the Mississippi River, played a crucial role in the Union's victory. By gaining control of the Mississippi, the Union split the Confederacy in two, severely disrupting their ability to move troops and supplies. Union naval forces, under the command of Admiral David Farragut, also secured key victories in the Gulf of Mexico, further weakening the South's strategic position.

4. Emancipation Proclamation: President Abraham Lincoln's issuance of the Emancipation Proclamation in 1863 played a significant role in shaping the Union's goals and war aims. While primarily a political and moral decision, it also had a military impact by depriving the Confederacy of much-needed labor. It also allowed African Americans to join the Union army, bolstering their forces and providing additional manpower.

5. Leadership: Effective leadership was crucial for both sides, but the Union had several notable military leaders who contributed to their success. Generals such as Ulysses S. Grant, William Tecumseh Sherman, and Philip Sheridan played key roles in implementing successful strategies and winning crucial battles.

6. Industrial and Technological Advantages: The North's superior industrial capacity and technological advancements, such as railroads, telegraphs, and mass-produced weaponry, gave them a significant advantage. The North had a greater ability to produce and supply troops with essential resources, weapons, and equipment, contributing to their overall success.

It's important to note that the war was complex, and various factors ultimately contributed to the Union's victory. These strategies, among others, helped shape the outcome and paved the way for the preservation of the Union and the abolition of slavery.

Diplomacy played a limited role in the American Civil War, and its effectiveness was somewhat limited. There were several attempts made by both the Union and the Confederacy to engage in diplomatic efforts, but they did not lead to a resolution or end the conflict. Here are a few key aspects to consider:

The Confederacy sought international recognition as a sovereign nation, primarily from European powers like Britain and France, to gain economic and military support. However, despite their efforts, the European powers did not recognize the Confederacy as a separate entity. While some European nations sympathized with the South and even contemplated intervening, concerns over slavery, economic interests, and political stability ultimately prevented official recognition.

In 1861, the Union intercepted and seized the British mail steamer Trent, which carried Confederate envoys James Mason and John Slidell. This action nearly sparked an international incident, as Britain demanded the release of the envoys and threatened war. Eventually, the crisis was resolved diplomatically when the Union released the envoys, avoiding a full-scale conflict with Britain.

The Union attempted to prevent European powers from recognizing the Confederacy by emphasizing the abolitionist cause and their commitment to ending slavery. They aimed to present the Civil War as a moral struggle against slavery, hoping to deter international support for the Confederacy. These efforts were partially successful, as anti-slavery sentiment prevailed in Europe, making it difficult for the Confederacy to gain official recognition.

The Confederacy sent envoys to Europe, seeking recognition, support, and arms supplies. They hoped that European powers, in need of Southern cotton, would recognize their independence and intervene to break the Union blockade. Although European countries were interested in obtaining cotton, their concerns over sectionalism, potential repercussions from the United States, and a desire to maintain neutrality prevented official recognition.

In summary, while diplomacy had some impact during the Civil War, it did not lead to a significant resolution of the conflict or tip the scales decisively in favor of either side. Ultimately, the resolution came through military means and the defeat of the Confederacy on the battlefield.

Human rights played a significant role in diplomacy during the American Civil War, particularly in shaping international perceptions and influencing diplomatic decisions. Here are a few ways in which human rights considerations were important in diplomacy during this period:

The issue of slavery lay at the heart of the Civil War, and it had significant implications for diplomacy. The Union sought to portray the war as a moral struggle against the institution of slavery, aiming to gain international support and deter recognition of the Confederacy. By emphasizing the abolitionist cause, the Union aimed to appeal to the humanitarian sentiments and anti-slavery sentiment prevailing in many nations.

The Union's diplomatic efforts focused on highlighting their commitment to ending slavery and promoting human rights. They sought to generate moral pressure on European powers, painting the Confederacy as a bastion of slavery and oppression. By framing the war as a fight for human rights and freedom, the Union hoped to sway international public opinion and discourage support for the Confederacy.

Britain had a strong anti-slavery movement, and many British citizens were deeply opposed to the institution of slavery. These sentiments influenced public opinion and political discourse, making it difficult for the British government to officially recognize the Confederacy. Humanitarian concerns over the treatment of enslaved individuals in the American South made it politically challenging for Britain to support the Confederacy openly.

14
FREEDOM AND RECONSTRUCTION

The issuance of the Emancipation Proclamation in 1863 by President Abraham Lincoln had both domestic and diplomatic implications. While primarily a strategic and moral decision, the Proclamation aimed to disrupt the Southern economy by depriving the Confederacy of free labor. Internationally, it further solidified the Union's commitment to ending slavery and reaffirmed their pursuit of human rights. It also allowed African Americans to join the Union army, presenting the Union's cause as one of freedom and equality.

European powers, including Britain and France, were hesitant to officially recognize the Confederacy due to concerns over supporting a government built upon slavery. While economic interests and other strategic considerations also played a role, the moral dimension of the conflict and the Union's focus on human rights had an impact on the international stance towards the Civil War.

In summary, human rights considerations, particularly the issue of slavery, were central in shaping diplomatic efforts during the Civil War. The Union's emphasis on the abolitionist cause and their portrayal of the conflict as a fight for human rights played a role in influencing international perceptions and diplomatic decisions.

Following the end of the Civil War, the focus shifted towards Reconstruction, which aimed to rebuild the Southern states and reintegrate them into the Union. Reconstruction included efforts to establish civil rights for newly freed African Americans, ensure loyalty to the Union, and encourage economic recovery in the devastated regions.

Under Reconstruction policies, the Southern states were required to acknowledge the Thirteenth, Fourteenth, and Fifteenth Amendments to the United States Constitution, which abolished slavery, granted equal protection under the law, and established voting rights for African American men. Former Confederate states also had to draft new state constitutions and apply for readmission to the Union.

While the Reconstruction policies did impose certain conditions on the Southern states and institute changes in governance, they did not include any specific reparations payments. The focus was on rebuilding the Southern economy and society rather than seeking monetary compensation from the defeated Confederacy.

It's important to note that Reconstruction was a complex and often contentious period in American history, with ongoing debates over how to reconcile the nation and address the legacy of slavery. The process was eventually marked by the end of Reconstruction in 1877, leading to the resurgence of white supremacist policies and the abandonment of some of the gains made by African Americans during that time.

During Reconstruction (1865-1877), the federal government implemented various policies and initiatives aimed at rebuilding the Southern states after the Civil War. Reconstruction sought to address the issues of social, political, and economic transformation in the aftermath of slavery and secession. The ratification of the Thirteenth, Fourteenth, and Fifteenth Amendments abolished slavery, granted citizenship and equal protection under the law to all citizens (including former slaves), and protected voting rights regardless of race or color.

Congress passed a series of Reconstruction Acts in 1867 that divided the South into military districts, with the aim of enforcing these amendments and guaranteeing civil rights. These acts required former Confederate states to draft new state constitutions, grant suffrage to African American men, and apply for readmission to the Union.

The Freedmen's Bureau, created in 1865, provided assistance to newly freed slaves and poor white citizens. It established schools, hospitals, distributed food, mediated labor disputes, and helped former slaves secure land and employment.

The Civil Rights Act of 1866 passed by Congress, provided legal protection to African Americans by guaranteeing equal rights in courts and public accommodations. It aimed to counteract the Black Codes, which were discriminatory laws passed by Southern states to restrict the rights and freedoms of African Americans.

Under Reconstruction, the federal government appointed military governors, known as Reconstruction governments, to oversee the reestablishment of civil order and the development of new state governments in the Southern states.

In an attempt to address issues of land ownership and economic independence, efforts were made to redistribute land to former slaves and provide them with access to resources and tools. However, these efforts were largely unsuccessful, and many former slaves became trapped in exploitative sharecropping arrangements.

Reconstruction faced resistance and challenges, including the rise of white supremacist groups like the Ku Klux Klan, violent intimidation towards African Americans, and the controversial presidential election of 1876. Eventually, Reconstruction policies and protections were abandoned, leading to the end of Reconstruction in 1877.

The effectiveness and success of Reconstruction varied across different regions and communities. While some progress was made in establishing civil rights for African Americans and rebuilding the South, the end of Reconstruction saw the imposition of Jim Crow laws and the disenfranchisement of African Americans, leading to a long period of racial segregation and discrimination.

Abraham Lincoln, the 16th President of the United States, was shot on the evening of April 14, 1865. Lincoln was attending a play at Ford's

Theatre in Washington, D.C., when he was assassinated by John Wilkes Booth, a Confederate sympathizer and actor. The gunshot wound proved fatal, and Lincoln passed away the following morning on April 15, 1865.

Abraham Lincoln was shot by John Wilkes Booth, a Confederate sympathizer and actor, on the evening of April 14, 1865, at Ford's Theatre in Washington, D.C. Booth had vehemently opposed Lincoln's presidency and the Union cause during the Civil War. The assassination can be attributed to Booth's deep-seated hatred for Lincoln and his desire to aid the Confederacy.

Booth believed that killing Lincoln would cripple the Union government and allow the Confederacy to continue fighting. He saw Lincoln as a symbol of the Union cause and viewed his death as a way to strike a significant blow against the Union's leadership. Booth also had a personal vendetta against Lincoln due to his strongly anti-slavery stance and his speeches advocating for civil rights for African Americans.

Booth and his co-conspirators initially planned to kill not only Lincoln but also Vice President Andrew Johnson and Secretary of State William Seward as part of a broader conspiracy to destabilize the government. However, Booth was the only conspirator who successfully carried out their plan. After shooting Lincoln, Booth escaped from the theater but was later captured and killed by Union soldiers in Virginia.

After John Wilkes Booth shot President Abraham Lincoln on April 14, 1865, he fled the scene and went into hiding. Booth was a well-known stage actor and used his knowledge of the theater to evade capture for several days.

Booth escaped from Ford's Theatre, where he shot Lincoln, and made his way to southern Maryland. He was pursued by Union soldiers and eventually found refuge on a farm owned by Richard Garrett. However, federal troops tracked him down, surrounding the farm on April 26, 1865.

During the standoff, the soldiers set the barn where Booth was hiding on fire in an attempt to force him out. Booth refused to surrender and was shot in the neck by Sergeant Boston Corbett. He was then dragged out of the burning barn, paralyzed from the neck down.

Booth was brought to the porch of the Garrett farmhouse, where he lived for a few more hours before succumbing to his injuries. He died in the early hours of April 26, 1865.

After Booth's death, his body was identified and taken to Washington, D.C. for an autopsy and further identification. Eventually, he was buried in an unmarked grave at the Old Penitentiary Cemetery in Washington's Arsenal Penitentiary (now the site of the Lincoln Cottage).

Several of Booth's alleged co-conspirators in the assassination plot were apprehended, and many were sentenced to imprisonment or execution. This included Mary Surratt, who owned the boardinghouse where the conspirators met, and was the first woman to be executed by the United States government.

After Abraham Lincoln's assassination, Reconstruction continued under President Andrew Johnson, who assumed office in April 1865. Johnson's approach to Reconstruction differed significantly from that of Lincoln. While Lincoln had favored a more lenient plan towards the Southern states, Johnson's policies were seen as more sympathetic to the South and lenient towards former Confederates.

Johnson issued multiple pardons to former Confederates and allowed Southern states to quickly rejoin the Union without many of the protections for formerly enslaved African Americans that were advocated by Republicans in Congress. This approach clashed with the views of many Radical Republicans in Congress who believed that the South needed a more thorough reconstruction to ensure civil rights for freed slaves.

The Radical Republicans used their majority in Congress to pass several Reconstruction Acts in 1867 and 1868, which imposed military rule on the Southern states and required them to draft new constitutions protecting the rights of African

Americans. These acts also gave African American men the right to vote. Additionally, Congress passed the Fourteenth Amendment, granting citizenship and equal protection under the law to all individuals born or naturalized in the United States, regardless of race.

Throughout this period, conflicts and tensions between Johnson and Congress grew. Johnson vetoed many of the Reconstruction Acts, but Congress overrode his veto on several occasions. Congress also attempted to impeach Johnson in 1868, but he narrowly avoided removal from office.

Andrew Johnson, the 17th President of the United States, was almost impeached due to his clashes with Congress over Reconstruction and his attempts to block legislation aimed at protecting the rights of freed slaves.

Johnson became president in 1865 after the assassination of Abraham Lincoln. As a Southern Democrat and former slave owner from Tennessee,

Johnson's views on Reconstruction often clashed with the Republican-dominated Congress.

One of the main reasons Johnson faced impeachment was his opposition to the Radical Republicans' plans for Reconstruction. These Republicans sought to implement a more punitive approach toward the Southern states during the post-Civil War period and ensure the civil and political rights of freed slaves. However, Johnson consistently undermined their efforts.

One significant event that triggered the impeachment process was Johnson's dismissal of Edwin M. Stanton, the Secretary of War, in violation of the Tenure of Office Act. The act was passed by Congress to restrict the president's ability to remove certain officeholders without Senate approval. Johnson's removal of Stanton was seen as a direct challenge to Congress's authority and an attempt to block Reconstruction measures.

The House of Representatives voted to impeach Johnson on eleven articles of impeachment, with the main charges being related to his violation of the Tenure of Office Act and his obstruction of Reconstruction policies. The case then moved to the Senate for trial.

In the Senate trial, Johnson narrowly avoided conviction and removal from office. He was acquitted by just one vote, falling short of the necessary two-thirds majority needed for removal.

While Johnson was not impeached and removed from office, his presidency was significantly weakened by the process. He remained in office until the end of his term in 1869 but was unable to effectively govern or achieve his policy goals.

Reconstruction officially ended in 1877 after the contentious presidential election between Rutherford B. Hayes and Samuel Tilden. As part of the Compromise of 1877, Hayes was declared the winner, and in exchange, federal troops were withdrawn from the South, effectively ending the era of Reconstruction. With the withdrawal of federal troops, Southern states were left to revert to their pre-Reconstruction practices, and many civil rights gains made during Reconstruction were eroded.

The war debts associated with the Civil War were primarily owed by the federal government of the United States, rather than specific states or regions. The Union government, led by the northern states, incurred significant debt in financing the war effort. To fund the war, the Union government relied on various methods, including increased taxation, issuing government bonds, and borrowing from domestic and foreign sources. The total debt accumulated by the Union government during the Civil War was approximately $2.7 billion.

The major components of this war debt included:

1. Government Bonds: The federal government issued bonds to raise funds for the war. These bonds were essentially loans to the government with a promise of repayment over a specified period, typically with interest.

2. Treasury Notes: The Union also issued treasury notes, which were short-term debt instruments that acted as a form of currency. They were used to finance day-to-day expenses and were typically redeemed within a few years.

3. Greenbacks: To help finance the war, the Union government issued paper currency known as "greenbacks." These were not backed by a specific commodity or gold standard but were declared legal tender for all debts, public and private.

4. Loans: The Union government borrowed money from foreign governments, such as Britain, and private banks in the form of loans to finance the war effort.

After the Civil War, the federal government embarked on a plan to repay its war debt. This involved implementing various financial policies, such as increasing taxes and reducing government spending. The debt was eventually paid off through a combination of economic growth, fiscal discipline, and financial management.

15
SEGREGATION

It's worth noting that the Confederate states, which were the southern secessionist states that formed the Confederacy, also accrued debt to finance their war effort. However, their ability to borrow and repay was much more limited due to economic blockade, lack of international recognition, and ultimately the defeat of the Confederate army. The exact extent of Confederate war debts is difficult to determine, as many records were lost or destroyed during and after the war.

Unfortunately, racial tensions did not cease after the Civil War. While the war resulted in the abolition of slavery and the legal recognition of equal rights for African Americans through the Reconstruction Amendments (Thirteenth, Fourteenth, and Fifteenth), the end of the war did not bring about an immediate end to racism and discrimination.

During the Reconstruction era, African Americans achieved significant political and social gains, including the election of African American officials and the establishment of schools and institutions for former slaves. However, these gains were short-lived as racial divisions and violence began to reemerge.

The Southern states implemented a system of laws known as "Jim Crow" laws, which enforced racial segregation and denied African Americans basic civil rights. These laws, along with widespread violence and intimidation from white supremacist groups like the Ku Klux Klan, had a profound impact on African Americans' lives and perpetuated racial tensions.

Racial tensions persisted throughout the 20th century, with the continued existence of segregated schools, public facilities, and neighborhoods. African Americans faced systemic racism and discrimination in employment, housing, education, and access to public services.

The civil rights movement of the 1950s and 1960s sought to address these racial tensions and fight for equality. Civil rights activists, including Martin Luther King Jr., Rosa Parks, and many others, organized marches, sit-ins, and protests to challenge segregation and demand equal rights and opportunities for African Americans.

While substantial progress has been made in the fight against racism, racial tensions and discrimination remain issues in society today. Achieving equality and eradicating racial tensions continues to be an ongoing struggle.

Joe Biden's stance on segregation has been a subject of scrutiny and criticism, particularly regarding his early political career. In the 1970s, Biden was a senator from Delaware, a state with a deeply segregated history. During that time, he faced criticism for his positions on issues related to racial integration and school busing.

Biden opposed federally mandated busing as a means to desegregate schools, arguing that it was a flawed approach that could lead to more racial division. His stance aligned with many other politicians, including some within the civil rights movement, who believed that busing was an intrusive and ineffective method for achieving desegregation.

It is important to consider the historical context in which these debates occurred. The issue of school busing was highly contentious, and opinions on the best approach to achieve racial equality varied widely. Biden's position reflected a broader tension between desegregation goals and concerns about the impact of busing on communities.

However, it is crucial to note that Biden's position on segregation and busing evolved over time. In subsequent years, Biden has acknowledged that his stance on busing was not entirely in line with his overall commitment to civil rights. Throughout his career, including his tenure as vice president under Barack Obama, Biden has advocated for policies aimed at advancing racial equality and social justice.

It is worth mentioning that political figures' positions on historical issues can be complex and evolve over time. While Biden's earlier stance on segregation and busing has been criticized, it is essential to consider his entire record, including his subsequent actions and policy positions, when assessing his commitment to racial equality.

The return to Africa initiative, also known as the Back-to-Africa movement or colonization movement, began in the early 19th century. The American Colonization Society (ACS) was the primary organization behind the initiative, and it was founded in 1816. The ACS aimed to encourage the emigration of free African Americans and formerly enslaved individuals to Africa, specifically to the colony established in West Africa, known as Liberia.

The ACS believed that sending African Americans back to Africa would be a solution to racial tensions and slavery in the United States. They argued that African Americans would have better opportunities and prospects for success in Africa, and they sought to establish a community in Liberia that would serve as a model for African American self-governance.

The return to Africa initiative gained traction throughout the 19th century, with thousands of African Americans voluntarily emigrating to Liberia. However, it faced criticism from both abolitionists and African Americans who saw it as an attempt to solve the problem of slavery by removing African Americans from American soil rather than addressing their rights and equality at home.

The Back-to-Africa movement continued to have an impact into the early 20th century, but its prominence began to decline as the African American civil rights movement gained momentum in the United States. Today, Liberia remains a nation with historical ties to the return to Africa movement.

President James Monroe played a significant role in the early years of the American Colonization Society (ACS), which was founded in 1816. While not a founder of the ACS, Monroe became a prominent supporter and advocate for the organization during his presidency.

President Monroe offered his public support to the ACS, seeing it as a potential solution to the challenges of slavery and racial tensions in the United States. In his 1824 annual message to Congress, Monroe expressed support for the colonization of free African Americans in Africa and praised the ACS's efforts. He believed that the colonization movement could help safeguard the rights and welfare of African Americans while promoting racial harmony in the United States.

Furthermore, Monroe encouraged Congress to provide financial support to the ACS and facilitate the emigration of African Americans to colonies in Africa. In 1821, Congress appropriated $100,000 to the ACS to assist in establishing the colony of Liberia in West Africa.

James Monroe's endorsement and backing of the ACS helped lend credibility and momentum to the colonization movement. His support, along with that of other prominent figures, helped raise funds and mobilize support for the initiative. However, it is important to note that Monroe's views on colonization evolved over time, and he eventually shifted his focus towards the gradual emancipation of enslaved individuals within the United States.

After the Civil War, the idea of some African Americans returning to Africa gained some traction, particularly among a group called the "Back-to-Africa" movement. This movement was led by figures like Marcus Garvey, who advocated for the repatriation of African Americans to Africa. However, the vast majority of African Americans chose to remain in the United States instead of returning to Africa.

There are several reasons why many African Americans did not choose to return to Africa. Many African Americans had been born and raised in the United States, and their families had been in the country for generations. They had established communities, built relationships, and had ties to the land and culture. For them, America was their home, and they wanted to stay and fight for their rights in the country they considered theirs.

The end of the Civil War brought about significant political and social changes for African Americans. The passage of the Thirteenth, Fourteenth, and Fifteenth Amendments abolished slavery, granted citizenship, and protected voting rights for African American men. These milestones represented progress towards equality within the United States, and many African Americans believed in fighting for their rights in the country they were living in, rather than abandoning their efforts and returning to Africa.

Despite the challenges and discrimination they faced, many African Americans saw potential economic opportunities in the United States. The post-war period brought some changes to the Southern economy, with Reconstruction policies creating new possibilities for landownership and employment opportunities. Additionally, African Americans who had gained skills during slavery and the war sought to use those skills to build better lives for themselves and their families in America.

African Americans had created a unique cultural identity in the United States, blending African, European, and American influences. This unique cultural heritage had its roots in the experiences of enslavement, resistance, and survival. By staying in the United States, African Americans could continue to shape and contribute to this evolving culture, preserving their history and heritage while striving for equal rights and opportunities.

It is important to note that while most African Americans did not choose to return to Africa, there were some who did pursue repatriation, particularly through organizations like the American Colonization Society. These groups facilitated the migration of a small number of African Americans to locations such as Liberia and Sierra Leone in West Africa.

After the end of the American Civil War, there were some individuals from the Confederate

States who chose to go to the United Kingdom, but it was not a widespread phenomenon. Many prominent Confederate leaders and military officers faced the possibility of being tried for treason and chose to leave the United States. Some of them, such as Confederate President Jefferson Davis, sought refuge in countries like the United Kingdom to avoid prosecution.

However, it is important to note that the United Kingdom did not officially recognize or support the Confederate States during the Civil War. While there were some sympathizers in Britain, including those in the upper class who sympathized with the Confederacy's aristocratic heritage, the British government maintained a neutral stance throughout the war.

16
THE SOUTHERN PROMISE

After the Civil War, some former Confederates sought to escape the harsh conditions in the defeated South, including economic devastation and the disruption caused by Reconstruction policies. The United Kingdom and other countries offered these individuals a chance to start anew with potentially better economic prospects.

In addition to the United Kingdom, some former Confederates also chose to settle in other locations overseas, such as Mexico, Canada, and various countries in South America. These individuals often looked for opportunities to rebuild their lives and find work in these new environments. However, it should be noted that the number of Confederates who chose to relocate to foreign countries was relatively small compared to the overall population of the former Confederate States.

There were numerous factors that influenced the outcome of the Civil War, and it is challenging to pinpoint a single factor that could have completely changed the outcome. However, if one factor had been different, it could have potentially altered the course of the war. One significant factor that comes to mind is foreign intervention.

If a foreign power, particularly a European nation, had intervened on behalf of the Confederacy, it could have significantly shifted the balance of power and potentially led to Confederate victory. The Confederacy sought foreign recognition and support, especially from European powers like Britain and France. A decision by either of these nations to intervene militarily or provide significant military aid to the Confederacy could have proved decisive.

Foreign intervention would have put the Union at a severe disadvantage, as it would have faced a two-front war and additional logistical challenges. It might have weakened the Union's blockade of Confederate ports, allowing the Confederacy to more easily import supplies and export its agricultural products. This could have prolonged the war and potentially persuaded the Union to negotiate a peace treaty rather than achieve a military victory.

However, it's important to note that foreign intervention on behalf of the Confederacy was unlikely. European powers considered multiple factors, including the Confederacy's dependence on slavery and issues of international law, which led most countries to adopt a policy of neutrality during the American Civil War. Ultimately, the lack of significant foreign intervention allowed the Union to maintain its blockade and military pressure on the Confederacy, contributing to the Union's eventual victory.

In the immediate aftermath of the Civil War, the South did not embrace desegregation as a means to their advantage. Rather, the Southern states embarked on a period known as Reconstruction, which aimed to address the political, social, and economic consequences of the war. During this time, there were efforts to implement policies that would ensure political rights and civil liberties for African Americans.

However, as Reconstruction policies were rolled back and the South entered a period known as the Jim Crow era, the concept of desegregation was actively used by the Southern states to preserve white supremacy and maintain racial hierarchy. The primary objective of the Southern states during this period was to enforce racial segregation and limit the rights and opportunities of African Americans.

The South implemented laws commonly ireferred to as "Jim Crow laws." These laws, which encompassed various areas of life including education, transportation, public spaces, voting, and employment, mandated racial segregation and ensured the systematic subjugation and disenfranchisement of African Americans. The goal was to enforce racial separation and maintain white dominance.

In certain instances, Southern states strategically used desegregation to their advantage by manipulating the implementation of federal desegregation initiatives. For example, when faced with the Supreme Court's ruling in Brown v. Board of Education (1954) that segregation in public schools was unconstitutional, some Southern states adopted tactics such as "token desegregation" or "slow Southern resistance." These methods aimed to comply with the court's decision while maintaining significant racial segregation and delaying substantial integration efforts.

It's important to note that the use of desegregation by the South during this period was not aimed at achieving racial equality or justice, but rather at perpetuating racial discrimination and preserving white supremacy. The efforts to enforce segregation were eventually challenged and dismantled through the Civil Rights Movement, which aimed to secure equal rights and opportunities for African Americans and put an end to racial segregation and discrimination in the United States.

The removal of Confederate statues has been a topic of significant debate and controversy in recent years. The primary reasons behind the push to take down these statues are rooted in concerns about racism, historical memory, and a reassessment of the symbols and narratives that these statues represent. Many Confederate statues were erected during periods of racial tension and were intended to honor and memorialize Confederate leaders and soldiers. However, these statues often serve as symbols of white supremacy, as they were frequently installed as part of efforts to maintain racial segregation and intimidate African Americans during the era of Jim Crow laws. For many, these statues represent a painful reminder of the era of slavery and oppression, perpetuating a narrative of oppression and inequality.

Critics argue that Confederate statues propagate a distorted and romanticized version of history. These statues often portray Confederate leaders and soldiers as noble heroes fighting for a noble cause, downplaying or ignoring the central role of slavery as the primary catalyst for the Civil

War. Many advocates for removal argue for a more accurate and inclusive interpretation of history, one that acknowledges the true motivations and consequences of the Confederacy.

Confederate statues are predominantly located in public spaces, symbolically granting visibility and honor to a cause that is associated with racism and oppression. Many argue that these statues are an affront to communities that are diverse and aspire to values of equality and inclusivity. They contend that public spaces should reflect the shared values and aspirations of a community, rather than glorifying figures tied to a history of racial discrimination.

Removing Confederate statues is seen as a step towards acknowledging and redressing the deep wounds of racial inequality and historical injustices. For many, the presence of these statues reinforces and perpetuates divisions, making it difficult to move forward as a united society. Supporters of their removal argue that addressing these symbols can foster reconciliation, healing, and a more inclusive understanding of history.

It is important to note that opinions regarding the removal of Confederate statues vary. Some argue that these statues serve as historical markers and remind society of the complexities of history, while others view them as symbols of hate and oppression. The debates surrounding the removal of these statues reflect a wider societal discussion about the significance of historical memory, representation, and the fight against racism.

Numerous Confederate statues and monuments have been removed across the United States since the discussion around their presence and symbolism gained traction. The specific statues that have been taken down can vary from city to city and state to state. Here are a few notable examples:

1. Robert E. Lee Monument in Richmond, Virginia: The largest and most prominent Confederate statue in the country, the statue of Confederate General Robert E. Lee was removed from its pedestal on Monument Avenue in Richmond,

Virginia, on September 8, 2021, after a lengthy legal battle.

2. Stonewall Jackson Statue in Richmond, Virginia: The statue of Confederate General Stonewall Jackson was also located on Monument Avenue in Richmond. It was removed on July 1, 2020, following protests and calls for its removal.

3. Jefferson Davis Monument in Richmond, Virginia: The monument to Confederate President Jefferson Davis was located on Monument Avenue in Richmond. It was removed on June 10, 2020, after Governor Ralph Northam ordered its removal.

4. Confederate Soldiers and Sailors Monument in Richmond, Virginia: Located on Libby Hill in Richmond, this monument was removed on July 8, 2020, after the city's mayor and governor ordered its removal.

5. Silent Sam at the University of North Carolina at Chapel Hill: The Confederate monument known as "Silent Sam" was removed in August 2018 after

years of controversy and protests. It had stood on the university's campus since its dedication in 1913.

Please note that this list is not exhaustive, and there have been numerous other Confederate statues and monuments removed or relocated across the United States in recent years. The removal of these statues has been a contentious and complex process, often involving legal battles, public outcry, and civic debates.

The desire to preserve Confederate heritage in the South is rooted in a combination of factors, including historical pride, cultural identity, and a sense of regional autonomy. Some Southerners view Confederate heritage as an important part of their region's history and take pride in the valor and sacrifices of their ancestors who fought for the Confederacy during the Civil War. They may see Confederate symbols and monuments as a way to honor and commemorate their heritage and the memory of their ancestors.

Confederate symbols and heritage are deeply entrenched in Southern culture for many individuals. They may view the preservation of

Confederate symbols as central to their identity, linking them to a specific historical period, regional traditions, and a sense of belonging to the South. Southern heritage activists argue that the Confederate cause was largely about states' rights and autonomy. They argue that the preservation of Confederate symbols and monuments is a way to defend the importance of local governance and limit federal intervention in regional affairs.

Some proponents argue that removing or altering Confederate symbols erases or distorts history. They believe that these symbols provide an opportunity to remember and learn from the past, even if it includes acknowledging the problematic aspects of that history. Those who want to preserve Confederate heritage may see efforts to remove or critique Confederate symbols as part of a wider phenomenon of "political correctness" or "cancel culture." They argue that removing these symbols infringes upon free speech and the right to commemorate and express one's heritage.

It is essential to note that the perspectives on Confederate heritage are diverse within the

South, and not all residents support the preservation of Confederate symbols. Many argue that these symbols represent a painful legacy of slavery and racism, and their removal is necessary for inclusivity, healing, and moving toward a more equitable future. The preservation of Confederate symbols continues to be a contentious and complex issue, with ongoing debates and discussions across Southern states and beyond.

While it's impossible to predict the future with certainty, the likelihood of another civil war in the United States is generally considered to be extremely low. The United States has a robust system of democratic governance and strong institutions that serve as checks and balances. These institutions help maintain stability and resolve conflicts through peaceful means, such as the courts, legislature, and elections.

The United States is a diverse country with a wide range of issues that divide its citizens. The ideological divide between conservatives and liberals is a significant source of division. Disagreements on topics such as the role of government, taxation, healthcare, and social issues can be highly polarizing.

The debate over gun control is highly contentious. Advocates for stricter regulations believe it is necessary to reduce gun violence, while opponents argue that it infringes upon their Second Amendment rights.

Immigration policy, including topics such as border control, pathways to citizenship, and the treatment of undocumented immigrants, is a deeply divisive issue. Different perspectives exist on how to address immigration challenges and balance national security and humanitarian concerns.

The United States continues to grapple with racial inequality, which sparks heated discussions on issues such as police brutality, systemic racism, affirmative action, and economic disparities. These issues often lead to extensive disagreements on the best approaches to address and rectify these inequalities.

The topic of climate change and its relationship to policy decisions divides the country. While many recognize the urgency of addressing climate change, opinions diverge on how extensively the government should intervene, the role of renewable energy, and the impact on the economy.

Issues concerning LGBTQ+ rights, including marriage equality and transgender rights, often lead to fierce debates among Americans. Disagreements stem from differing beliefs about personal freedom, religious beliefs, and the scope of civil rights protections.

The abortion debate centers on the legal, medical, ethical, and religious aspects of reproductive rights. Divergent views exist on when, how, and under what circumstances abortion should be permissible, leading to significant divisions.

These are just a few examples of the many issues that divide the United States. It is worth noting that the national discourse and divisions on these topics can change over time as societal norms evolve and new issues emerge.

Despite political divisions and societal conflicts, the United States still maintains a shared national identity that is tied to the ideals of freedom, democracy, and the Constitution. Many Americans value the unity of the country and its democratic principles over individual divisions.

Unlike the situation leading up to the American Civil War in the 19th century, the United States today is characterized by a more interconnected society with increased cross-regional movement, cultural exchange, and shared experiences. This has contributed to a greater sense of national identity and reduced the regional tensions that were more prevalent in the 1860s.

The United States is heavily interconnected economically, with trade networks and shared economic interests across state lines. This interdependence serves as a deterrent to conflict, as it would have severe economic consequences for all parties involved.

The American legal system and respect for the rule of law play significant roles in preventing violence and ensuring disputes are resolved peacefully. The judiciary is available to address constitutional issues, and the majority of Americans still have trust in the legal system's ability to administer justice.

While political polarization and social divisions exist in the United States, the vast majority of Americans still prefer peaceful means of resolving conflicts and advancing their interests. The focus tends to be on engaging in political discourse, elections, advocacy, and peaceful protests rather than resorting to violence.

The desire of some individuals and communities in Southern states to keep the Confederacy relevant today can be attributed to a variety of reasons. It is important to note that perspectives on this issue vary among individuals, and not all residents of Southern states have the same views. Here are some potential reasons for the desire to maintain Confederate relevance:

Some individuals may have a deep sense of pride in their regional history and the valor of their ancestors who fought for the Confederacy during the Civil War. They may view Confederate symbols and heritage as a way to honor and commemorate that history.

For many Southerners, Confederate symbols and heritage are closely intertwined with their cultural identity. They may see the preservation of these symbols as necessary to hold on to a distinct regional identity that connects them to their heritage, traditions, and sense of belonging to the South.

Some proponents argue that removing or diminishing the relevance of the Confederacy is an attempt to erase or revise history. They may believe that preserving Confederate symbols provides an opportunity to remember and learn from the past, even if it means acknowledging the problematic aspects of that history.

Some individuals may have personal connections to the Confederacy through their family history or ancestors who served in the Confederate military. They may feel a personal connection and a desire to honor their family's legacy.

The desire to maintain Confederate relevance may also stem from opposition to what some perceive as "political correctness" or "cancel culture." They argue that removing these symbols infringes upon their right to express their cultural, historical, or regional heritage and see it as an attack on free speech.

It is important to recognize that the perspectives on Confederate relevance are diverse within Southern states. Many people argue that Confederate symbols represent a painful legacy of slavery and racism and emphasize the need to move towards a more inclusive and equitable future. The discussion around Confederate symbolism and heritage is complex and often contentious, with ongoing debates and conversations taking place at local, state, and national levels.

The reasons behind the continuation of Confederate pride and the exhibition of Confederate symbols in Southern states vary from person to person and region to region. While not all residents of Southern states share the same views, many individuals and communities in the Southern states view Confederate pride as a way to honor their ancestors who fought in the Civil War. They may see displaying Confederate symbols as a way to connect with their regional history and maintain a sense of pride in their heritage.

Confederate symbols and heritage are often intertwined with the cultural identity of many Southerners. Some argue that the display of these symbols is a way to preserve a unique Southern identity and celebrate the distinct traditions, values, and history of the region.

Opposition to removing Confederate symbols can be rooted in resistance to what some perceive as erasure of history or a changing societal and cultural landscape. They may argue that preserving these symbols is important to remember and learn from the past, regardless of the problematic aspects associated with the Confederacy.

17
HATE GROUPS

Confederate symbols and pride can also be seen as a unifying factor for some individuals within the Southern states, fostering a sense of community and belonging. Confederate pride can also be intertwined with political ideologies. Some individuals may view the display of Confederate symbols as a means of expressing their political beliefs or opposing certain policy initiatives.

It is important to note that the topic of Confederate pride is deeply divisive and controversial. Opponents argue that Confederate symbols represent a painful legacy of slavery, racism, and oppression, and that their display is offensive and hurtful to marginalized communities. The perspective on Confederate symbolism and heritage varies widely, and ongoing debates and discussions continue to shape the public discourse on this issue.

The Ku Klux Klan (KKK) and the Confederacy are related in the sense that they both have ties to the history of racial discrimination and white supremacy in the United States. However, it is important to understand that the KKK is an extremist, white supremacist organization that emerged long after the Confederate era, and their relationship is not direct or officially connected. The Confederate States of America (CSA) was a breakaway nation formed by Southern states during the Civil War (1861-1865), which sought to preserve slavery and an agrarian Southern way of life. The Confederacy's historical significance lies in its role as a belligerent in the Civil War.

Following the Civil War, during the Reconstruction era (1865-1877), the South underwent a period of political, social, and economic transformation as the Confederate states were reintegrated into the United States. During this time, racial tensions remained high, and African Americans faced significant challenges to their newly gained civil rights.

The first incarnation of the Ku Klux Klan was formed in 1865 immediately after the end of the Civil War. It was predominantly composed of white Southerners who sought to resist and undermine the social, political, and economic advancements of African Americans during Reconstruction.

The KKK's primary objectives were to restore white supremacy, eradicate the political and social power of African Americans, and maintain the racial status quo of the pre-war South. They used intimidation, violence, and acts of terror to suppress African American civil rights and prevent social and political progress. Over time, the KKK began to adopt and appropriate Confederate symbols, such as the Confederate battle flag, as part of their imagery and identity. This association is based on a shared belief in racial hierarchy and white supremacy rather than any official connection between the two.

It is crucial to note that while some individuals who identify with Confederate heritage may have personal views overlapping with those of the KKK, it does not mean that all people who express an attachment to Confederate symbols or history endorse or support the KKK's ideology or actions. The relationship between the KKK and the Confederacy is nuanced and not universally representative of all individuals who identify with Confederate heritage or history. Historically, there have been elected officials in the United States who were affiliated with or supported the Ku Klux Klan (KKK). However, providing an exact number of KKK members in Congress now or historically is challenging for several reasons.

Firstly, membership in the KKK has been secretive and its membership records have not been publicly disclosed. Secondly, individuals affiliated with the KKK might not openly declare their membership or association due to the negative public perception and legal consequences that could follow. Moreover, personal viewpoints and affiliations can change over time, and accurately attributing KKK membership can be difficult.

It is important to note that white supremacist or extremist beliefs and affiliations can exist outside of the KKK and may influence individuals in politics. However, it is not appropriate to assume that all elected officials who held or currently hold Confederate sympathies or are associated with Confederate symbols necessarily have direct ties to the KKK or hold white supremacist views.

Neo-Nazis and Confederate groups are different entities, but they can share some ideological commonalities centered around white supremacy and racial discrimination. It is important to understand the distinctions between these two groups:

Neo-Nazis are individuals or organizations that adhere to neo-Nazi ideology, which is a modern iteration of the Nazi ideology that emerged in Germany during the 20th century. Neo-Nazis embrace Adolf Hitler's racist beliefs, anti-Semitism, and the concept of Aryan racial superiority. They often advocate for the establishment of a white ethnostate and use symbols, slogans, and rhetoric associated with Nazism.

Conversely, Confederate groups often gather around the historical symbolism and heritage of the Confederate States of America (CSA) during the American Civil War. Some groups might express pride in Confederate history, culture, or share an attachment to symbols associated with the Confederate States, such as Confederate flags or monuments. While not all individuals or groups affiliated with Confederate heritage adopt white supremacist beliefs, there have been instances of overlap or shared sympathies between Confederate ideologies and white supremacist viewpoints.

While there may be overlaps and commonalities between some individuals or factions within these groups, it's essential to recognize that not all Confederate group members are neo-Nazis, and not all neo-Nazis identify with or express an attachment to Confederate symbolism or history. Both groups represent different historical contexts and movements, even though they may share similar white supremacist ideologies.

The United States operates under a democratic system with a representative form of government where power is dispersed across multiple branches and levels. While it is not accurate to say that a single group or set of groups "controls" America, various entities and institutions exert influence on the nation. Here are some key entities that play significant roles:

The U.S. government consists of three separate branches: the executive, legislative, and judicial. They have distinct powers and responsibilities to provide checks and balances on one another. The government is designed to be accountable to the people through elections and strives to represent the interests of the American population through democratic processes.

The United States has a multi-party system dominated by two major political parties: the Democratic Party and the Republican Party. These parties field candidates for elections and play a crucial role in shaping policy agendas, mobilizing voters, and influencing legislation at the national, state, and local levels.

Interest groups represent specific causes, industries, or demographic groups and aim to influence public policy outcomes. They advocate on behalf of their members or constituents by lobbying lawmakers, mobilizing public opinion, and engaging in various forms of advocacy.

Corporations hold significant economic power and often play influential roles in shaping policies and public debates. Various business groups, such as chambers of commerce and trade associations, represent the interests of businesses and industries in the policymaking process.

Media outlets, including newspapers, television networks, online platforms, and radio stations, play a vital role in informing the public, shaping public opinion, and holding those in power accountable. Different media organizations have varying levels of influence and biases.

Grassroots movements and activism have historically played a significant role in shaping American society. From civil rights movements to environmental advocacy to social justice initiatives, individuals and groups outside traditional power structures have mobilized to effect change and influence the political landscape.

It is important to note that power dynamics in American society can be complex and constantly evolving, with influence and control dispersed among various entities. The relative importance and influence of each group can vary depending on the issue, context, and public sentiment.

David Duke's relationship with Donald Trump has been a source of concern and controversy for several reasons:

David Duke is a known white supremacist and a former grand wizard of the Ku Klux Klan (KKK). He has a long history of promoting racist and anti-Semitic ideologies. His association with Trump raises concerns because it suggests a level of alignment or sympathy with white supremacist beliefs.

During the 2016 presidential campaign, when asked about David Duke's endorsement,

Trump initially offered unclear answers, deflecting and disavowing Duke multiple times but without explicitly rejecting his support. This lack of a strong and immediate repudiation raised concerns about Trump's willingness to distance himself from white supremacists.

Duke's endorsement and his associations with Trump were seen by some as indicative of broader support for Trump from elements within the alt-right and white nationalist movements. While it is important to note that not all Trump supporters are white supremacists, the alignment of prominent white supremacist figures with his campaign raised alarms about the potential normalization of their ideologies.

Trump's presidency was marked by significant racial tensions and controversies. Critics argued that his rhetoric and policies, such as his remarks about Mexicans, his response to white supremacist violence in Charlottesville, and his approach to immigration, emboldened and energized extremist elements within society.

It is important to note that Trump has disavowed Duke's support on multiple occasions, including issuing explicit condemnations of white supremacists and disavowing any association with Duke during his presidency. However, concerns persist due to the initial response and the broader context of racial tensions during his time in office.

The relationship between Duke and Trump is troubling because it raises questions about racial tolerance, inclusivity, and the potential influence of white supremacist ideologies within political discourse and support structures. It highlights the ongoing need for vigilance in addressing and confronting racism and extremism in society.

18
A DESIRE FOR STRONG LEADERSHIP

Both domestic race relations and foreign relations are crucial, but they are distinct and interconnected issues that require attention and action.

Domestic race relations are essential because they shape the social fabric of a nation and affect the well-being and opportunities of its citizens. Racial inequality and social injustice within a country can lead to divisions, tensions, and systemic disadvantages for minority communities. Building inclusive societies that value diversity, promote equality, and address racial disparities is vital for the overall progress and harmony of a nation.

On the other hand, foreign relations involve a country's interactions and diplomatic relationships with other nations. These relationships impact various aspects, including political, economic, security, and cultural dimensions. Strong foreign relations are essential for promoting national interests, cooperating on global challenges, fostering peaceful resolutions to conflicts, and advancing international cooperation.

Both domestic race relations and foreign relations can have significant consequences for a country's stability, security, and overall prosperity. Neglecting either area can lead to problems and limitations.

It's important to note that the two are interconnected. How a country handles its internal racial issues can impact its image and credibility on the international stage. Similarly, international events and foreign policies can have both direct and indirect effects on domestic race relations, such as influencing perceptions, attitudes, and dynamics within the nation.

In conclusion, both domestic race relations and foreign relations are crucial to the well-being and success of a nation. It is crucial for governments and individuals to address and work towards improving both aspects in order to foster a just and inclusive society while also fostering positive relationships with other nations.

Some argue that during Trump's presidency, there were record-low unemployment rates, a strong stock market, and increased GDP growth. They believe that Trump's business background and policies, such as tax cuts and deregulation, contributed to a thriving economy.

Supporters of Trump believe that his tough stance on immigration and border security helped protect American jobs and national security. They highlight the construction of the border wall and efforts to enforce immigration laws as evidence of his commitment to this issue.

Trump's supporters argue that his "America First" policy emphasized prioritizing American interests in trade deals, international relations, and military engagements. They believe this approach protected American industries and workers.

Trump signed criminal justice reform legislation, known as the First Step Act, which aimed to reduce recidivism rates and address issues in the justice system. Some supporters appreciate this effort as a positive step towards criminal justice reform.

Trump's administration pursued deregulatory policies, with the goal of reducing bureaucratic red tape and fostering a business-friendly environment. Supporters argue that this approach encouraged business growth and job creation.

It is important to recognize that these points represent a specific perspective. In contrast, supporters of Joe Biden may present different arguments in favor of his presidency. An objective assessment of each candidate should examine their overall records, policy positions, leadership qualities, and the alignment of their values and priorities with your own.

Donald Trump was not reelected for a variety of factors that contributed to his loss in the 2020 presidential election. It is important to note that multiple factors and circumstances can influence the outcome of an election. The Trump administration faced criticism for its handling of the COVID-19 pandemic, including perceived delays in response, inconsistent messaging, and challenges in implementing an effective nationwide strategy. The impact of the pandemic on public health and the economy played a significant role in shaping voters' perceptions and priorities.

The 2020 election took place against a backdrop of significant political polarization in the United States. Trump's divisive rhetoric and policies throughout his tenure exacerbated political divisions, leading to a highly polarized electorate. While he maintained strong support from his base, his approach alienated some moderate and independent voters.

Instances of racial injustice and protests over police brutality, such as the killing of George Floyd, intensified during Trump's presidency. Critics argue that Trump's response to these events, including his rhetoric and handling of protests, failed to effectively address the issues and contributed to increased racial tensions and divisions.

Trump's unique persona and communication style, characterized by controversial tweets, blunt language, and confrontational debates, appealed to some voters. However, it also led to concerns about his temperament, credibility, and ability to unite the country. These factors influenced voter decisions, particularly among swing voters and independents.

The level of voter turnout and successful voter mobilization efforts by Joe Biden's campaign and Democratic organizations played a crucial role in the outcome. Biden's campaign was effective in mobilizing a broad coalition of supporters and had a strong presence in key battleground states, ultimately securing the necessary electoral votes.

Hillary Clinton played a pivotal role in helping Joe Biden and Barack Obama win their respective presidential elections. Here are some ways she supported their campaigns:

As a highly respected figure within the Democratic Party, Hillary Clinton's endorsements carried significant weight. During the 2008 presidential campaign, Clinton endorsed Obama after a hard-fought primary contest. Her endorsement helped to unify the party and rally her supporters behind Obama's candidacy. Similarly, in 2016, Clinton endorsed and campaigned for Biden, lending her support and credibility as a former Secretary of State and presidential candidate.

Clinton's extensive network, experience, and ability to connect with voters were valuable assets. In both election cycles, she actively campaigned for Biden and Obama, speaking at rallies, attending fundraisers, and engaging with voters across the country. Clinton's influence and popularity among Democratic donors also helped to mobilize substantial financial support for their campaigns.

Clinton's extensive experience in government and foreign policy positioned her as a valuable advisor to Biden and Obama. Her deep knowledge and insights on critical issues, such as healthcare, economic development, and international relations, helped shape their policy platforms and provided strategic guidance.

Clinton played a crucial role in promoting party unity and rallying Democrats behind the eventual nominees. In 2008, despite the competitive primary campaign with Obama, she delivered a unifying speech at the Democratic National Convention, urging her supporters to unite behind Obama's candidacy. Similarly, in 2016, Clinton's endorsement and active support for Biden helped to consolidate Democratic support and promote party cohesion.

Throughout her career, Clinton has been actively involved in mobilizing and engaging with voters. Her efforts to reach out to diverse communities, particularly women and minority voters, played a critical role in mobilizing support for Biden and Obama. Clinton's focus on gender equality and women's rights also resonated with many voters, particularly in the 2016 election.

While it is impossible to determine the precise impact of any individual on an election's outcome, Hillary Clinton's contributions were instrumental in helping Joe Biden and Barack Obama win their respective presidential campaigns.

Hillary Clinton's unsuccessful presidential campaigns can be attributed to a combination of factors. Firstly, she had a high level of name recognition and a long history in politics, which made her a deeply polarizing figure. She faced significant opposition and negative sentiment from those who were critical of her policy positions, her handling of controversies such as the private email server, and perceptions of her as part of the political establishment.

One major challenge Hillary Clinton faced was the perception of untrustworthiness and lack of transparency. The controversy surrounding her use of a private email server during her tenure as Secretary of State raised concerns about her handling of sensitive information and contributed to doubts about her integrity in the minds of some voters.

Hillary Clinton's campaign faced criticism for its messaging and strategic choices. Some argue that her campaign did not effectively address the concerns and frustrations of working-class voters, particularly in key battleground states. There were perceptions of a lack of a clear and inspiring vision or message that could differentiate Clinton from her opponents.

Despite winning the popular vote in 2016, Clinton fell short in the Electoral College, as the United States presidential election is ultimately determined by that system. The concentration of support in certain geographic regions and the importance of swing states played a significant role in her electoral defeat.

Hillary Clinton faced biases and unrealistic expectations as a female candidate for the presidency. She encountered gender-based scrutiny and double standards that affected perceptions of her competence and likability. These biases and challenges were significant factors in shaping voter attitudes and opinions.

It is crucial to note that winning a presidential campaign is a complex and multifaceted task. While these factors played a role in Hillary Clinton's unsuccessful presidential campaigns, analyzing the outcome of an election is a nuanced process that encompasses a wide range of variables and circumstances.

It is important to note that the outcome of an election is influenced by a combination of factors, including the candidates' policy positions, campaign strategies, voter perceptions, and the prevailing socio-political climate. The 2020 election reflected a combination of issues specific to that year, as well as broader trends and dynamics within American politics.

The events that took place on January 6, 2021, at the United States Capitol were undoubtedly alarming and marked a significant breach of security and democratic norms. While the actions of the rioters were unprecedented and shook the nation, it is important to note that they did not escalate into a full-blown civil war.

A civil war typically involves a protracted conflict between organized and armed factions within a country, often with differing political, ideological, or territorial goals. The events on January 6th, although deeply concerning and violent, did not reach the level of a civil war.

It is worth noting that American democracy faced a severe test on that day, as the peaceful transfer of power was disrupted by the breach of the Capitol building. However, the violence was contained, and order was eventually restored by law enforcement and the National Guard. The incident was widely condemned by a broad spectrum of political leaders.

While the events on January 6th exposed significant divisions and tensions within American society, the country did not descend into a widespread and prolonged armed conflict that would meet the technical definition of a civil war.

Nevertheless, the events of that day underscored the importance of upholding democratic processes, promoting unity, and addressing the underlying grievances and divisions within society. They serve as a reminder of the need for open dialogue, respect for the rule of law, and the peaceful resolution of political disputes.

Improving race relations in America is a complex and multifaceted task that requires ongoing efforts from individuals, communities, institutions, and the nation as a whole. While there is no single solution, here are some steps that can be taken to foster greater understanding, equality, and unity:

Foster historical and contemporary education about racial issues, including the history of racism, the experiences of different racial and ethnic groups, and the contributions they have made to American society. Encourage open conversations, dialogue, and sensitivity training to increase awareness and understanding.

Acknowledge and address systemic racism that exists in areas such as criminal justice, education, employment, and housing. Implement policies that promote equal opportunities and fair treatment for all individuals, regardless of their race or ethnicity.

Encourage diverse representation at all levels of leadership in government, business, and other institutions. By having diverse voices involved in decision-making processes, perspectives are enriched, and policies can be developed that address the needs and concerns of all communities.

Promote empathy and understanding by creating safe spaces for people of different backgrounds to share their experiences, listen to each other, and engage in constructive dialogue. This can be facilitated through community forums, town halls, and initiatives that encourage people to step out of their echo chambers and engage with those they may not typically interact with.

Encourage community-building activities that foster connections and positive interactions between people from different racial and ethnic backgrounds. This can include intercultural festivals, volunteer opportunities, neighborhood collaborations, and initiatives aimed at supporting local minority-owned businesses.

Promote diverse and inclusive media representation that challenges stereotypes and highlights the richness and diversity of different racial and ethnic groups. Encourage media outlets to provide accurate and fair coverage of racial issues and stories.

Advocate for policies and legislation that promote equality, equity, and social justice. This can include reforms in areas such as criminal justice, voting rights, housing, and education that address systemic inequalities and uplift marginalized communities.

It is important to recognize that meaningful change will take time and effort from individuals and society as a whole. Combining these strategies with a commitment to empathy, respect, and understanding can contribute to a more inclusive and racially equitable America.

Made in United States
Cleveland, OH
05 November 2026

47672909R00127